T0324701

Infinite Matrices and Their Recent Applications

P.N. Shivakumar • K.C. Sivakumar
Yang Zhang

Infinite Matrices and Their Recent Applications

P.N. Shivakumar
Department of Mathematics
University of Manitoba
Winnipeg, MB, Canada

K.C. Sivakumar
Department of Mathematics
Indian Institute of Technology, Madras
Chennai, TN, India

Yang Zhang
Department of Mathematics
University of Manitoba
Winnipeg, MB, Canada

ISBN 978-3-319-30179-2 ISBN 978-3-319-30180-8 (eBook)
DOI 10.1007/978-3-319-30180-8

Library of Congress Control Number: 2016933332

Printed on acid-free paper

This Springer imprint is published by Springer Nature
The registered company is Springer International Publishing AG Switzerland

Preface

Roughly speaking, an infinite matrix is a twofold table $A = (a_{i,j})$, $i, j \in \mathbb{N}$, of real or complex numbers. A general theory for infinite matrices commenced with Henri Poincaré in 1844. Helge von Koch carried on the quest after Poincaré, and by 1893, he had established "routine" theorems about infinite matrices. In 1906, David Hilbert applied infinite matrices to solve certain integral equations. This caught other mathematicians' eyes. Since then, numerous applications of infinite matrices appeared both in mathematics and in other sciences. In particular, in the mathematical formulation of physical problems and their solutions, infinite matrices stem more naturally than finite matrices.

Though there exist a rich literature and extensive texts on finite matrices, the only major work focusing on the general theory of infinite matrices, to the best of our knowledge, is the excellent text by Cooke [20]. Much of the material presented in that book reflects early stages of development. Bernkopf [12] studies infinite linear systems as prelude to Operator Theory. Since then, a lot of progress has been achieved in the theory and applications of infinite matrices, through the contributions of many mathematicians. The intention of this treatise is to introduce and also to present an in-depth review of some of the recent aspects of infinite matrices, together with some of their modern applications. The subject is abundant in engaging research problems, and it is our hope that the present monograph will motivate further research and applications. It may be mentioned here that the coverage of the material is by no means exhaustive and reflect the mathematical interests of the authors.

In the past, infinite matrices played an important role in the theory of summability of sequences and series, whereas in this monograph, we are mainly concerned with the theory of finite and infinite matrices, over the fields of real numbers, complex numbers, and over quaternions. We emphasize topics such as sections or truncations and their relationship to the linear operator theory on certain specific separable and sequence spaces. Most of the matrices considered in this monograph typically have special structures like being diagonally dominated, tridiagonal, and possess certain sign distributions and are frequently nonsingular. Such matrices arise, for instance, from solution methods for elliptic partial differential equations.

We focus on both theoretical and computational aspects concerning infinite linear algebraic equations, differential systems, and infinite linear programming, among other topics. Recent results on the non-uniqueness of the Moore-Penrose and group inverses are surveyed. Quaternions, well known as an extension of the complex number field, play an important role in contemporary mathematics. They have also been widely used in other areas like altitude control, 3D computer graphics, signal processing, and quantum mechanics. Here, efficient algorithms for computing $\{1\}$-inverses and Moore-Penrose inverses of matrices with quaternion polynomial entries are presented. New results on extensions of matrix classes like P-matrix, Q-matrix, and M-matrix to infinite dimensional spaces are surveyed. A finite dimensional approximation scheme for semi infinite linear programs is proposed, and its application to doubly infinite linear programs is considered.

Techniques like conformal mapping, iterations, and truncations are used to derive precise estimates in some cases and explicit lower and upper bounds for solutions of linear systems in the other cases. Topics such as Bessel's and Mathieu's equations, viscous fluid flow in doubly connected regions, digital circuit dynamics, and eigenvalues of the Laplacian are covered. The eigenvalues of the Laplacian provide an explanation for the various cases when the shape of a drum cannot be determined just by knowing its eigenvalues. To hear the shape of a drum is to infer information about the shape of the drumhead from the sound it makes, i.e., from the list of basic harmonics, via the use of mathematical theory. Other applications include zeros of a Taylor series satisfying an infinite Vandermonde system of equations and simultaneous flow of oil and gas.

Winnipeg, MB, Canada P.N. Shivakumar
Chennai, TN, India K.C. Sivakumar
Winnipeg, MB, Canada Yang Zhang

Acknowledgements

P.N. (Shiv) Shivakumar dedicates this book to the memory of his parents: Saroja and Professor Nagappa.

The authors wish to thank Drs. Joseph Williams, Safoura Zadeh, and Kanaka Durga.

This research was partially supported by the grants from the Natural Sciences and Engineering Research Council of Canada (NSERC) and the National Natural Science Foundation of China (NSFC 11571220).

Table of Contents

Chapter 1
Introduction

In this chapter, first we provide a brief overview of some of the advances that have been made recently in the theory of infinite matrices and their applications. Then we include a summary of the contents of each chapter.

Infinite matrices have applications in many branches of classical mathematics such as infinite quadratic forms, integral equations, and differential equations. As illustrated in Chap. 8, this topic has applications in other sciences besides mathematics, as well. A review of some of the topics of this monograph was recently undertaken by Shivakumar and Sivakumar [111]. In this monograph, apart from elaborating on some of the topics that are discussed in that review, we include other interesting topics such as quaternion matrices and infinite dimensional extensions of certain positivity classes of matrices.

Gaussian elimination, the familiar method for solving systems of finitely many linear equations in finitely many unknowns, has been around for over two hundred years. Unlike the matrix methods, matrices were not used in the early formulations of Gaussian elimination, until the mid-twentieth century.

The word "matrix" was coined by James Sylvester in 1850. Roughly speaking, a matrix over a field F is a two fold table of scalars each of which is a member of F. A simple example of an infinite matrix is the matrix representing the derivative operator which acts on the Taylor series of a function. In the seventeenth and eighteenth centuries, infinite matrices arose from attempts to solve differential equations using series methods. This method would lead to a system of infinitely many linear equations in infinitely many unknowns. Therefore, a main source of infinite matrices is solutions of differential equations.

As mentioned in the preface, a general theory for infinite matrices originated with Henri Poincare in 1844. Infinite determinants were first introduced into analysis in 1886 in the discussion of the well-known Hill's equation. By 1893, Helge Von Koch established standard theorems on infinite matrices. In 1906, David Hilbert attracted the attention of other mathematicians to the subject by solving a Fredholm integral equation using infinite matrices. Since then, many theorems, fundamental

P.N. Shivakumar et al., *Infinite Matrices and Their Recent Applications*,
DOI 10.1007/978-3-319-30180-8_1

to the theory of operators on function spaces were discovered although they were expressed in special matrix terms. In 1929 John Von Neumann showed that an abstract approach is more powerful and preferable rather than using infinite matrices as a tool to the study of operators. Hence, the modern operator theory stems from the theory of infinite matrices. Despite this, the infinite matrix theory remains a subject of interest for its numerous and natural appearances in mathematics as well as in other sciences. For example, in mathematical formulation of physical problems and their solutions, infinite matrices appear more naturally than finite matrices. Some of the recent applications include flow of sap in trees, leakage of electricity in coaxial cables (attenuation problem), cholesterol problem in arteries, and simultaneous flow of oil and gas.

Let us summarize the contents of this monograph. In Chap. 2, we consider the notion of diagonal dominance of complex matrices and discuss the various results that guarantee invertibility of such matrices. Possibly reducible matrices satisfying a chain condition are discussed. Recent results on specially structured matrices like tridiagonal matrices and matrices satisfying certain sign patterns are reviewed.

Given an infinite matrix $A = (a_{i,j})$, $i, j \in \mathbb{N}$, a space X of infinite sequences, and $x = (x_i)$, $i \in \mathbb{N}$, we define Ax by $(Ax)_i = \sum_{j=1}^{\infty} a_{ij}x_j$ provided this series converges for each $i \in \mathbb{N}$, and define the domain of A as $\{x \in X : Ax \text{ exists and } Ax \in X\}$. We define an eigenvalue of A to be any scalar λ for which $Ax = \lambda x$ for some nonzero x in the domain of A. A matrix A is diagonally dominant if

$$|a_{ii}| \geq \sum_{j \neq i} |a_{ij}|, \text{ for all } i.$$

Let $A = (a_{ij})$, $i, j \in \mathbb{N}$ be an infinite matrix. The linear differential system

$$\frac{d}{dt}x_i(t) = \sum_{j=1}^{\infty} a_{ij}x_j(t) + f_i(t), \; t \geq 0, \; x_i(0) = y_i, \; i = 1, 2, \ldots,$$

is of considerable theoretical and applied interest. In particular, such systems occur frequently in the theory of stochastic processes, perturbation theory of quantum mechanics, degradation of polynomials, infinite ladder network theory, etc. Arley and Brochsenius [3], Bellman [7], and Shaw [100] are the notable mathematicians who have studied this problem. In particular, if A is a bounded operator on l^1, then the convergence of a truncated system has been established. For instance, Shivakumar, Chew and Williams [118] provide an explicit error bound for such a truncation.

Diagonally dominant infinite matrices occur in solutions of elliptic partial differential equations as well as solutions of second order linear differential equations. The eigenvalue problem for particular infinite matrices including diagonally dominant matrices is studied by Shivakumar, Williams and Rudraiah [119]. A discussion of these topics is presented in Chap. 3.

In Chap. 4, first we provide a brief review of generalized inverses of matrices with real or complex entries followed by a discussion on the Moore–Penrose inverses of operators between Hilbert spaces. Certain non-uniqueness results for generalized inverses of infinite matrices are reviewed later.

Chapter 5 considers the case of generalized inverses of matrices whose entries are quaternionic polynomials. After a discussion on the theoretical aspects, some algorithmic approaches are proposed. The contents report recent results in this area.

A vast literature exists for M-matrices for which more than fifty characterizations are given. Relatively little is known for the case of infinite dimensional spaces. Chapter 6 presents a review of M-operator results obtained recently, including some results on extensions of two other matrix classes quite well known in the theory of linear complementarity problems.

Infinite linear programming problems are linear optimization problems where, in general, there are infinitely many (possibly uncountable) variables and constraints related linearly. There are many problems arising from the real world situations that can be modelled as infinite linear programs. Some examples include the bottleneck problem of Bellman in economics, infinite games, and continuous network flow problems [2]. A finite dimensional approximation scheme for semi-infinite linear programming problems is presented in Chap. 7, where an application to obtaining approximate solutions to doubly infinite programs is considered.

The importance of eigenvalue problems concerning the Laplacian is well documented in classical and modern literature. Finding the eigenvalues for various geometries of the domains has posed many challenges for solution methods, which have included infinite systems of algebraic equations, asymptotic methods, integral equations, and finite element methods. The eigenvalue problems of the Laplacian is represented by Helmholtz equations, Telegraph equations, or the equations of the vibrating membrane and is given by

$$\frac{\partial^2 u}{\partial x^2} + \frac{\partial^2 u}{\partial y^2} + \lambda^2 u = 0 \text{ in } D, \text{ and } u = 0 \text{ on } \partial D,$$

where D is a plane region bounded by smooth curve ∂D. The eigenvalues k_n and corresponding eigenfunctions u_n describe the natural modes of vibration of a membrane. The eigenvalues of the Laplacian provide an explanation for the various cases when the shape of a drum cannot be determined just by knowing its eigenvalues. To hear the shape of a drum is to infer information about the shape of the drumhead from the sound it makes, i.e., from the list of basic harmonics, via the use of mathematical theory. In 1964, John Milnor with the help of a result of Ernst Witt showed that there exist two Riemannian flat tori of dimension 16 with the same eigenvalues but different shapes. However, the problem in two dimensions remained open until 1992, when Gordon, Webb, and Wolpert showed the existence of a pair of regions in the plane with different shapes but identical eigenvalues. The regions are non-convex polygons. Chapter 8 provides more information on this intriguing problem, among other interesting applications.

Chapter 2
Finite Matrices and Their Nonsingularity

2.1 Introduction

Nonsingularity of matrices plays a vital role in the solution of linear systems, matrix computations, and numerical analysis. A large variety of problems arising in computational fluid mechanics, fluid dynamics, and material engineering that are modelled using difference equations or finite element methods require that the matrix under consideration is nonsingular, in order for the numerical schemes to be convergent. In spite of the large scale availability of excellent software for the computation of eigenvalues, there is always a growing need for new results on invertibility of matrices and inclusion regions of spectra of matrices. This is true especially due to the fact that in practical problems, matrices are dependent on parameters. Further, bounds for eigenvalues of finite matrices usually lead to derivation of bounds for the spectra of infinite matrices. Due to these reasons, discovering new sufficient conditions for matrix invertibility and eigenvalue inclusion regions are very relevant even today.

In this section, we give several different criteria for an $n \times n$ matrix $A = (a_{ij})$ to be nonsingular. We also present in most of these cases, estimates for the elements of $A^{-1}\left(= \dfrac{A_{ji}}{\det A}\right)$, where A_{ji} represents the cofactor of a_{ij} and $\det A$ is the determinant of A. In what follows $|.|$ stands for the modulus. Let us recall that a (finite) linear system

$$\sum_{j=1}^{n} a_{ij}x_j = b_i, \quad i = 1, 2, \ldots, n \tag{2.1}$$

could be succinctly written as

$$Ax = b,$$

P.N. Shivakumar et al., *Infinite Matrices and Their Recent Applications*,
DOI 10.1007/978-3-319-30180-8_2

where $A = (a_{ij})$ is the matrix obtained from the coefficients, $x \in \mathbb{R}^n$ is the column vector of the unknowns, and $b \in \mathbb{R}^n$ is the right-hand side (requirement) vector. Assuming that A^{-1}, (the inverse of A) exists, we then have $x = A^{-1}b$. Since $\det A \neq 0$ one has by Cramer's rule, the formula for the unknowns, given by:

$$x_j = \sum_{k=1}^{n} \frac{A_{kj}}{\det A} b_k, j = 1, 2, \ldots, n. \tag{2.2}$$

Invertibility is the central theme of this chapter and the material here is organized as follows. In the next section, we recall the notion of diagonally dominant matrices and present many results including some classical ones with an intention of providing a historical perspective. In Sect. 2.3, we present some recent results on the invertibility of a diagonally dominant matrix which is possibly reducible. A sufficient condition is given in terms of what we call as a chain condition. In the next section, viz., Sect. 2.4, results for tridiagonal matrices are presented. Certain improved bounds for the inverse elements are recalled here and bounds for the norm of the inverse matrix are given. The concluding section of this chapter presents invertibility results relating to two classes of matrices with given sign patterns on their entries.

We close this introductory part by recalling the notion of M-matrices. These matrices will be referred to in Sects. 2.2 and 2.3. A real matrix A of order $n \times n$ is called a *Z-matrix*, if the off-diagonal entries of A are nonpositive. It is easy to see that if A is a Z-matrix, then A could be written as $A = sI - B$, with $s \in \mathbb{R}$ and B is entrywise nonnegative, i.e., $B \geq 0$. A Z-matrix A with the representation as above is called an *M-matrix* if one has the inequality $s \geq \rho(B)$, where $\rho(B)$ denotes the spectral radius of B, namely the maximum of the moduli of the eigenvalues of B. A very frequently used and well-known result for an M-matrix A (with the representation as above) is that $s > \rho(B)$ if and only if A is invertible and that all the entries of the inverse of A are nonnegative, i.e., $A^{-1} \geq 0$. Matrices satisfying the last condition are called *inverse positive* matrices. Thus, an invertible M-matrix is inverse positive. There are more than fifty equivalent conditions for a real matrix to be an M-matrix. These are documented in the excellent book by Berman and Plemmons [11]. We shall also have the opportunity to present some very recent results on infinite M-matrices, albeit as particular cases of general results on operators over infinite dimensional spaces. These appear in Sect. 6.5 of Chap. 6.

2.2 Diagonal Dominance

In this section, we discuss the rather classical notion of diagonal dominance. For $A = (a_{ij}) \in \mathbb{C}^{n\times n}$, (with nonzero diagonals) $a_{ii} \neq 0$ for $1 \leq i \leq n$, suppose that the following hold:

$$\sigma_i |a_{ii}| = \sum_{j=1, j \neq i}^{n} |a_{ij}|, \ \sigma_i \geq 0, i = 1, 2, \ldots, n \tag{2.3}$$

We say that A is *row diagonally dominant*, if $\sigma_i \leq 1$ for all $i = 1, \ldots, n$. A is said to be *strictly row diagonally dominant*, if $\sigma_i < 1$ for all $i = 1, \ldots, n$. Similar definitions apply for columns. The significance of the notion of row (or column) diagonal dominance in the numerical solutions of partial differential equations is well documented (see for instance [131]). Schneider [99] mentions that it was one L. Levy who published the diagonal dominance theorem (for what we now call as Metzler matrices, i.e., negative Z-matrices) in the late nineteenth century, showing that a strictly diagonally dominant matrix is nonsingular. Desplanques later gave a proof for the general case. Hence this result is referred to as the Levy–Desplanques theorem. We shall not get into the historical details here, but merely point to the excellent account of Schneider [99].

There are a number of proofs showing that A is nonsingular, if A is strictly row diagonally dominant. Taussky credits a long list of contributors in [129]. We recall only the most prominent ones. Let $N = \{1, 2, \ldots, n\}$ and

$$J = \{i \in N : \ |a_{ii}| > \sum_{j \neq i} |a_{ij}|\}. \tag{2.4}$$

If $J = N$, then A is strictly row diagonally dominant. In this case, an application of the Gerschgorin circle theorem shows that the determinant of A does not vanish (pp. 106, [8] or Theorem 6.1.10, [44]). Interestingly, one could reverse this argument. The nonsingularity of a strictly diagonally dominant matrix implies Gerschgorin theorem (see Exc. 5, pp. 24, [131]). Proofs were also given by Ostrowski [82] and [83], Taussky (Theorem I, [129]), and Varga (Theorem 1.8, [131]). For instance, let us state the following result, by Ostrowski [83], which will be referred to a little later, in Sect. 2.4. He gave upper bounds for the inverse elements of a strictly row diagonally dominant matrix A, which we state, next. As before, let numbers σ_i, $i = 1, 2, \ldots, n$ be defined by

$$\sigma_i |a_{ii}| = \sum_{j \neq i} |a_{ij}|, \tag{2.5}$$

so that $\sigma_i < 1$, $i = 1, 2, \ldots, n$. Set $B = A^{-1}$. Using the usual convention of denoting $B = (b_{ij})$, one has the following bounds:

$$\frac{1}{|a_{jj}|(1 + \sigma_j)} \leq |b_{jj}| \leq \frac{1}{|a_{jj}|(1 - \sigma_j)}, \ j = 1, 2, \ldots, n.$$

$$|b_{ij}| \leq \sigma_j |b_{ii}|, \ i \neq j, \ i, j = 1, 2, \ldots, n.$$

Let us note that the entries of A and B, by definition, are related by the formula $b_{ij} = \frac{A_{ji}}{\det A}$ where, as mentioned in the beginning, A_{ji} is the cofactor of the entry a_{ij}. Observe also that, two sided bounds are presented for the diagonals of the inverse matrix, whereas only one-sided bounds are given for the off-diagonal elements. In Sect. 2.4, we shall review several results which present different lower bounds and also recall how the upper bounds could be improved for matrices with a special structure, viz., tridiagonal matrices.

It is noteworthy to observe that the proof (of the invertibility of a strictly diagonally dominant matrix), by Taussky referred to as above, is reproduced in some of the recent texts in linear algebra.

In the general case when J is not necessarily the whole of N, however, if A is irreducible, then A is nonsingular, so long as $J \neq \emptyset$. This was shown by Taussky (Theorem II, [129]). This is sometimes referred to as "Olga Taussky's theorem." Nowadays, the importance of these results in the proof of the convergence of Jacobi and Gauss-Seidel iterative methods is demonstrated even in a first course on numerical analysis. In subsequent discussions, we show how most of the extensions of diagonal dominance follow the trend that a certain strict diagonal dominance or a mere diagonal dominance together with an irreducibility assumption guarantees invertibility.

The infinity-norm condition number denoted by $\kappa_\infty(A)$ has an important role in numerical linear algebra. This is defined as

$$\kappa_\infty(A) = \| A \|_\infty \| A^{-1} \|_\infty.$$

Here $\| . \|_\infty$ is the operator norm induced by the vector norm $\| . \|_\infty$, where $\| v \|_{infty} := max\{|v_i| : I \in N\}$.

For example, let x^* denote an approximate solution of the linear equation $Ax = b$. Then the relative size of the residual is given by

$$\frac{\| Ax^* - b \|_\infty}{\| b \|_\infty}.$$

It then follows that the relative error of the approximate vector x^*, given by

$$\frac{\| x - x^* \|_\infty}{\| x \|_\infty}$$

is bounded by the product of the relative size of the residual and $\kappa_\infty(A)$. It is for this reason that bounds on the infinity norm of the inverse of a matrix A are important. In this connection, let us turn our attention to results on finding bounds on the infinite norm of the inverse of strictly row diagonally dominant matrices. Varah (Theorem 1, [130]) proved the following result. In the notation as above, set $\alpha = \min_k\{|a_{kk}(1 - \sigma_k)|\}$. Then

$$\|A^{-1}\|_\infty = \max_i \sum_{k=1}^{n} \frac{|A_{ki}|}{\det A} \leq \frac{1}{\alpha}. \tag{2.6}$$

A similar result gives an upper bound for the 1-norm of the inverse of a strictly column diagonally dominant matrix. Using his results, Varah obtained a lower bound for the smallest singular value of a matrix A which has the property that both A and A^* are strictly row diagonally dominant (Corollary 2, [130]). Let us observe that in the particular case of a strictly row diagonally dominant matrix A, where all the diagonal entries are equal to 1, by writing $A = I - B$, one has $\| B \|_\infty < 1$ and so Varah's result reduces to the classical result

$$\| (I - B)^{-1} \|_\infty \leq \frac{1}{1 - \| B \|_\infty},$$

which is proved even in a first course in numerical analysis and functional analysis.

Next, we recall a certain result of Varga obtained as a generalization of the result of Varah in the previous paragraph. For $A \in \mathbb{C}^{n \times n}$, define its *comparison matrix* by $(m_{ij}) = M_A \in \mathbb{R}^{n \times n}$:

$$m_{ii} := |a_{ii}| \text{ and } m_{ij} := -|a_{ij}|, \quad i \neq j, \quad i, j \in N.$$

Then M_A is a Z-matrix and coincides with A if A is a Z-matrix with nonnegative diagonal elements. A will be referred to as a *H-matrix* if M_A is a nonsingular M-matrix (so that ${M_A}^{-1} \geq 0$, as mentioned in the introduction). In such a case A is also nonsingular and one has the following relationship, which is attributed to Ostrowski (see also [44]).

$$|A^{-1}| \leq {M_A}^{-1}.$$

Here, $|B|$ denotes the matrix each of whose entries is the modulus of the corresponding entry of the matrix B. Varga attributes the definition of a H-matrix to Ostrowski. A little later, we include some recent results on H-matrices.

Proceeding with the general discussion, for an H-matrix A, define

$$U_A := \{u > 0 : M_A u > 0, \ \|u\|_\infty = 1\},$$

where $\|v\|_\infty = max\{|v_i| : i \in N\}$ and for $w \in \mathbb{R}^n$, we use the notation $w > 0$ to denote the fact that all the entries of w are positive; $w \geq 0$ will denote that all the entries of w are nonnegative. It follows that $U_A \neq \emptyset$. Let e denote the vector all of whose entries are 1. Define

$$u^* := \frac{{M_A}^{-1} e}{\|{M_A}^{-1} e\|_\infty}$$

and

$$f_A(u) := \min\{(M_A u)_i : i \in N, u \in U_A\}.$$

Then $f_A(u) > 0$ and in particular, one has

$$f_A(u^*) := \max\{f_A(u) : u \in cl(U_A)\},$$

where $cl(.)$ denotes the topological closure. Varga shows that (Lemma 1 and Theorem 1, [133])

$$\|M_A{}^{-1}\|_\infty = \frac{1}{f_A(u^*)} \geq \|A^{-1}\|_\infty.$$

A lower bound for the smallest singular value of A is also obtained (Theorem 2, [133]).

In what follows, we include some recently obtained results for H-matrices. Define

$$R_i = \sum_{j=1, j \neq i}^{n} |a_{ij}|, \ 1 \leq i \leq n.$$

Denote

$$N_1 := \{i \in N : 0 < |a_{ii}| \leq R_i\}$$

and its complement by

$$N_2 := \{i \in N : |a_{ii}| > R_i\}.$$

Suppose that both these sets are nonempty. Then a sufficient condition for a matrix to be a H-matrix is proved by Gan and Huang (Theorem 1, [35]). Suppose that for every $i \in N_1$ the following inequalities hold:

$$|a_{ii}| > \frac{\alpha_i}{|a_{ii}|}\{\sum_{k \in N_1, k \neq i} \frac{|a_{kk}|}{\alpha_k}|a_{ik}| + \sum_{k \in N_2} \frac{\alpha_k}{|a_{kk}|}|a_{ik}|\},$$

where α_k are certain positive constants. Then A is an H-matrix. Note that strict inequalities are required to hold for indices from the index set N_1. However, suppose that at least one inequality is strict in N_1 and not necessarily for all the indices, then as in the case of the usual diagonal dominance, the irreducibility of A guarantees that A is an H-matrix (Theorem 2). Another sufficient condition is presented in Theorem 4 and a necessary condition is given in Theorem 5. All these statements involve complicated inequalities and are not included here. We refer to [35] for the precise details.

Huang and Xu proposed an extension of strict row diagonal dominance in (Definition 1, [48]) and called it α-strict (row) diagonal dominance for $\alpha \in [0, 1]$. We do not present the details here, but merely point out that they obtain results similar to the strict row diagonal dominant matrices and show, for instance (Theorem 1) that an α-strict (row) diagonally dominant matrix must be an H-matrix. In the spirit of results discussed earlier, they show that a certain α-diagonally dominant matrix which is also irreducible must be an H-matrix.

Let us also include an interesting generalization of Varah's result, involving two matrices. Let A be an $n \times n$ strictly row diagonally dominant matrix and B be an $n \times m$ matrix, both with complex entries. Yong (Lemma 2.2, [139]) shows that the following inequality holds:

$$\|A^{-1}B\|_\infty \leq \max_i \frac{\sum_{j=1}^m |b_{ij}|}{|a_{ii}| - \sum_{j \neq i} |a_{ij}|}.$$

When $m = n$ and $B = I$, one has Varah's inequality, at once.

Before moving on to the next set of ideas, let us turn our attention to another generalization of strict diagonal dominance as proposed by Beauwens [6]. The author calls $A \in \mathbb{C}^{n \times n}$ as *lower semistrictly diagonally dominant* if A is row diagonally dominant and satisfies the following inequalities:

$$|a_{ii}| > \sum_{j=1, j \neq i}^n |a_{ij}|, \; 1 \leq i \leq n.$$

A is called *semistrictly diagonally dominant* if there is a permutation matrix Q such that QAQ^T is lower semistrictly diagonally dominant. The author shows that this notion provides a way of separating the properties of diagonally dominant matrices which depend on irreducibility from those which do not. It is shown that (Theorem 2.1, [6]) A is irreducibly diagonally dominant if and only if it is semistrictly diagonally dominant and irreducible. A diagonally dominant Z-matrix with nonnegative diagonal entries is an invertible M-matrix if and only if it is semistrictly diagonally dominant (Theorem 3.1, [6]).

As mentioned earlier, an immediate proof of the nonsingularity of a strictly row diagonally dominant matrix is provided by the Gerschgorin circle theorem. Another well-known theorem for eigenvalue localization is the theorem on the ovals of Cassini.

Zhang and Gu [142] considered a condition weaker than diagonal dominance and whose geometric interpretation concerns the location of the origin in relation to the ovals of Cassini. Li and Tsatsomeros [66] later referred to such matrices as doubly diagonally dominant. In what follows, we take a brief look at these results. As before, for $A \in \mathbb{C}^{n \times n}$, let

$$R_i = \sum_{k=1, k \neq i}^n |a_{ik}|, \; 1 \leq i \leq n.$$

A is said to be *doubly diagonally dominant*, if

$$|a_{ii}||a_{jj}| \geq R_i R_j, \ i,j \in N, \ i \neq j.$$

If the inequalities are strict for all distinct $i,j \in N$, then A is called *strictly doubly diagonally dominant*. On the other hand, if A satisfies the inequalities as above with at least one inequality holding strictly, and is irreducible, then A is called *irreducibly doubly diagonally dominant*. Let us denote by **G**, the set of doubly diagonally dominant matrices, by $\mathbf{G_1}$, the set of strictly doubly diagonally dominant matrices and by $\mathbf{G_2}$, the set of irreducibly doubly diagonally dominant matrices.

We develop some terminology before stating the main result of Zhang and Gu [142]. One can associate a *directed graph* $\Gamma(A)$ with A as follows: The vertices of $\Gamma(A)$ are $1, 2, \ldots, n$. There is a directed arc from vertex i to vertex j for every $a_{ij} \neq 0, \ i \neq j$. Clearly this directed graph depends only on the off-diagonal entries of A and has no loops. A *circuit* in the graph is a sequence of distinct vertices $i_1, i_2, \ldots, i_{p+1}$ with $i_{p+1} = i_1$ and each pair of vertices i_j, i_{j+1} is an arc. Then Brualdi (Theorem 2.9, [17]) shows the following: Suppose that A is irreducible and

$$\prod_{i \in \gamma} |a_{ii}| \geq \prod_{i \in \gamma} R_i(A)$$

for all circuits γ in $\Gamma(A)$, with strict inequality for at least one circuit. Then A is nonsingular.

It is noteworthy that Brualdi unified and generalized several classical results concerning inclusion regions and estimates for the eigenvalues of matrices using the notion of directed graphs. Let us now turn our attention to the result of Zhang and Gu [142], referred to, as above. Zhang and Gu show that (Theorem 1, [142]) if A is irreducible and satisfy

$$|a_{ii}||a_{jj}| \geq R_i R_j, \ i \neq j, \ i,j \in \gamma$$

for every circuit γ in $\Gamma(A)$, with strict inequality for at least one circuit, then A is nonsingular.

Li and Tsatsomeros [66] study further properties of doubly diagonally dominant matrices. They observe the following: An analogue of the Levy–Desplanques theorem holds, viz., that if $A \in \mathbf{G_1}$, then A is nonsingular. If $A \in \mathbf{G_2}$, then it does not imply that A is an H-matrix, even not nonsingular, either. If $A \in \mathbf{G_2}$, then the comparison matrix M_A is an M-matrix (possibly nonsingular). A is an H-matrix if and only if M_A is nonsingular. Among other things, they show that the Schur complement of a doubly diagonally dominant matrix inherits this property, which, however, does not inherit strict double diagonal dominance. They describe a situation when this holds, too. We do not include the details here.

We mention briefly, the work of Tam, Yang and Zhang [128], who obtained results that generalize, strengthen, and provide clarifications on the work of Brualdi, Zhang, and Gu. The main results are new sufficient conditions for invertibility of an irreducible complex matrix. Statements on eigenvalues and eigenvectors that lie on the boundary of spectrum inclusions of an irreducible matrix are also made.

Let us also take a look at the results of Farid [30]. Let $A \in \mathbb{C}^{n \times n}$ be diagonally dominant and let us denote the rows of A by $a^1, a^2, \ldots, a^n$. Suppose that there exist $k \in N$ and complex numbers $z_1, z_2, \ldots, z_n$ with $z_k = 1$ such that

$$a^k = \sum_{j=1, j \neq k}^{n} z_j a^j.$$

Then one has $|a_{kk}| = R_k$, where R_k is the sum of the moduli of all the entries of the kth row excluding the diagonal entry, defined as before. One also has the implication

$$m \in J \Longrightarrow z_m = 0,$$

in the representation for a^k given above. On the other hand, if $z_m = 0$ for some $m \in N$, then $a_{im} = 0$ whenever $z_i \neq 0$. In particular, $a_{km} = 0$. This is proved in (Theorem 2.1, [30]). This result, along with certain properties of diagonally dominant singular matrices, is used to establish a criterion for the nonsingularity of a diagonally dominant matrix with nonzero diagonal entries, in (Theorem 4.1, [30]). The statement of this result is too technical to be included here.

Farid, in [31], establishes relationships between some classes of matrices with properties that are variations of diagonal dominance. Sufficient conditions are given in order for a matrix to satisfy such generalized diagonal dominance properties. The details are not included here.

Let us make a brief mention of an excellent unified approach to diagonal dominance taken by Kostic [58]. Here, the author introduces the notion of diagonally dominant-type (DD-type) matrices and shows that the maximal nonsingular DD-type class is the class of (nonsingular) H-matrices. A new nonsingularity result that defines a DD-type class of matrices is proved. This is motivated by the fact that, using an infinity norm in the context of strictly diagonally dominant matrix may not be appropriate and so some other p-norm could be used, instead. Subsequently, the author discusses an equivalence principle for eigenvalue localizations, monotonicity, compactness, isolation, inertia, and spectral radius principles. New upper bounds are obtained for the spectral radius.

Let us next look at a certain invertibility condition which, however, has no relevance to diagonal dominance. This refers to a certain result of Gil [36] and is just a sample of the many results on invertibility and nonnegative invertibility that are available in the literature. Our intention in including it in this section comes from the fact that the matrices are invertible (in fact, inverse positive) and that an upper bound for the inverse is given. We give the precise statement of the main result. Let A be a complex square matrix of order n all of whose diagonal entries are nonzero. Define

$$\alpha_k := max_{1 \leq j \leq k-1} |a_{jk}|, \; 2 \leq k \leq n$$

and

$$\beta_k := max_{k+1 \le j \le n} |a_{jk}|, \; 1 \le k \le n-1$$

using which one defines the quantities below:

$$g_u := \prod_{k=2}^{n} (1 + \frac{\alpha_k}{|a_{kk}|})$$

and

$$g_l := \prod_{k=1}^{n-1} (1 + \frac{\alpha_k}{|a_{kk}|}).$$

Finally, define

$$\theta := (g_u - 1)(g_l - 1).$$

Then Gil (Theorem 2.1, [36]) shows that if $\theta < 1$, then A is invertible and one has

$$\| A^{-1} \|_\infty \le \frac{g_u g_l}{\alpha_0 (1-\theta)},$$

where $\alpha_0 > 0$ is the minimum of the moduli of the diagonal entries of A. It follows that such a matrix A must be inverse positive, i.e., $A^{-1} \ge 0$. Note that the matrix A above, in general, is neither diagonally dominant nor is an M-matrix. Let us also give a reference to another work where a whole of class of symmetric circulant matrices and symmetric pentadiagonal Toeplitz matrices with positive inverses have been identified. These are the results (Theorems 1 and 2) of Meek [72] and again, the matrices are neither diagonally dominant nor are M-matrices. We refer the reader to the citation for details.

In this concluding paragraph, we present some results which use strict diagonal dominance in certain other areas of mathematics. As before, given $A \in \mathbb{C}^{n \times n}$, let numbers σ_i, $i = 1, 2, \ldots, n$ be defined by

$$\sigma_i |a_{ii}| = \sum_{j \ne i} |a_{ij}|. \tag{2.7}$$

Let A be strictly row diagonally dominant so that $\sigma_i < 1$. Define

$$l_{kk} = |a_{kk}| - \sum_{j>k} \sigma_j |a_{kj}|$$

and

$$u_{kk} = |a_{kk}| + \sum_{j>k} \sigma_j |a_{kj}|$$

with $l_{nn} = u_{nn} = |a_{nn}|$. Then Brenner [15] has shown that one has

$$l_{kk} \leq |det\, A| \leq |u_{kk}|.$$

In what follows, we briefly present two interesting results on the bounds of the values of certain important polynomials, obtained as an application of the bounds as above. Recall that the Chebychev polynomial T_n is defined by the formula

$$T_n(x) := cos(n\, cos^{-1} x),\ x \in \mathbb{R}$$

and satisfies the three-term recurrence relation

$$T_n(x) := 2xT_{n-1}(x) - T_{n-2}(x),$$

with $T_0(x) = 1$ and $T_{-1}(x) = 0$. Now, let $C_n \in \mathbb{C}^{n \times n}$ be the matrix, each of whose super diagonal and sub diagonal entries equals 1, whose first $n-1$ diagonals are equal to $2x$ with the last diagonal equalling x, $x \in \mathbb{R}$ being fixed. Observe that $det\, C_n = T_n(x)$, $x \in \mathbb{R}$. Then C_n is strictly row diagonally dominant if and only if $|x| > 1$. By computing the lower and upper bounds for $det\, C_n$ (in place of the matrix A) similar to the inequalities above, one has the following result: If $|x| > 1$ and $n \geq 2$, then (Theorem 4.05, [16])

$$\frac{(2|x|^2 - 1)^{n-1}}{|x|^{n-2}} \leq |T_n(x)| \leq \frac{(2|x|^2 + 1)^{n-1}}{|x|^{n-2}}.$$

These results complement the well-known results for estimating T_n when $|x| \leq 1$.

Let us include bounds for one more polynomial, namely the Legendre polynomials. These are given by the formula:

$$P_n(x) := \frac{1}{2^n n!} \frac{d^n}{dx^n}(x^2 - 1)^n,\ x \in \mathbb{R}.$$

P_n satisfies the recurrence relation

$$nP_n(x) = (2n-1)xP_{n-1}(x) - (n-1)P_{n-2}(x),$$

with $P_0(x) = 1$ and $P_{-1}(x) = 0$, for all x. This time, let $F_n \in \mathbb{C}^{n \times n}$ be the matrix, whose super diagonal and sub diagonal entries are the numbers $1 - \frac{1}{n}, 1 - \frac{1}{n-1}, \ldots, 1 - \frac{1}{2}$ in that order and whose diagonals are $x(2 - \frac{1}{n}), x(2 - \frac{1}{n-1}), \ldots, x(2-1)$ in that order, with $x \in \mathbb{R}$ being fixed. Then $det\, F_n = P_n(x)$, $x \in \mathbb{R}$ and F_n is strictly row diagonally dominant if and only if $|x| > \frac{4}{3}$. It is shown (Theorem 6.02, [16]), that when $n \geq 2$ and for $|x| > \frac{4}{3}$ one has

$$|x| \prod_{j=2}^{n} \{(2 - \frac{1}{j})|x| - (1 - \frac{1}{j})^{\frac{1}{2}}\} \leq |P_n(x)| \leq |x| \prod_{j=2}^{n} \{(2 - \frac{1}{j})|x| + (1 - \frac{1}{j})^{\frac{1}{2}}\}.$$

2.3 A Chain Condition

In this section, let us recall another subclass of matrices still remaining in the class of row diagonally dominant matrices. Any matrix in this class turns out to be nonsingular, once again.

Suppose that the $n \times n$ matrix $A = (a_{ij})$ is row diagonally dominant and $J \neq \emptyset$. Assume that, there exists for each $i \notin J$, a sequence of nonzero elements of the form $a_{i,i_1}, a_{i,i_2}, \ldots, a_{i,i_k}$ with $i_k \in J$. Then A is nonsingular. This was established by Shivakumar and Chew (Theorem, [103]). The proof of this statement uses an interesting result by Fan [28], who has shown that for a matrix $A = (a_{ij})$ and an M-matrix $B = (b_{ij})$ whose diagonal entries are related by the inequalities

$$|b_{ii}| \leq |a_{ii}| \text{ for all } i \in N$$

and whose off-diagonal elements satisfy the inequalities

$$|a_{ij}| \leq |b_{ij}| \text{ for } i \neq j,\ i,j \in N,$$

the inequality $\det A \geq \det B$ holds. We may observe that the nonsingularity result stated above extends to cases where A (satisfies the chain condition and) is possibly reducible.

An interesting special case is obtained if one considers the class of Z-matrices all of whose diagonals are positive, satisfying the chain condition as above. It then follows that (Corollary 4, [103]) A is nonsingular and that $A^{-1} \geq 0$, viz., A is an invertible M-matrix.

In a subsequent work, the same authors studied an iterative procedure for numerically solving linear systems. Let $A = (a_{ij})$ be row diagonally dominant and $J \neq \emptyset$. Suppose further that the chain condition described above holds. We seek to solve numerically, the linear system $Ax = b$. Let A be decomposed as

$$A = D(I + L + U),$$

where D is the diagonal part of A, L is the strictly lower part of A, U is the strictly upper part of A, and I is the identity matrix. Let M be defined as

$$M = (1 + \alpha\omega L)^{-1}\{(1 - \omega)I - (1 - \alpha)\omega L - \omega U\}$$

for $0 < \omega < 1$ and $0 \leq \alpha \leq 1$. In (Theorem 2, [104]) the authors show that $\rho(M) < 1$, thereby showing that the stationary iterative method $x^{(k+1)} = Mx^{(k)} + g$ is convergent to the unique solution of the equation $Ax = b$.

Let us now consider a generalization of the result of Shivakumar and Chew, discussed above. This was achieved by Varga [132]. A matrix $(b_{ij}) = B \in \mathbb{R}^{n \times n}$ is of *generalized positive type* if there exists $u \in \mathbb{R}^n$ such that

$$u > 0,\ Bu \geq 0 \text{ and } \{i \in N : (Bu)_i > 0\} \neq \emptyset$$

and further satisfies the following chain condition: for every $i_0 \in N$ with $(Bu)_{i_0} = 0$, there exist indices $i_1, i_2, \ldots, i_r \in N$ with $b_{i_k i_{k+1}} \neq 0,\ 0 \leq k \leq r-1$ such that $(Bu)_{i_r} > 0$. It is now easy to observe that this idea generalizes both the strict (row or column) diagonal dominance and irreducible diagonal dominance.

Let us recall that M_A denotes the comparison matrix of any given $A \in \mathbb{C}^{n\times n}$. Given A, let $\Omega(A)$ denote the set of all matrices B which have the property that the corresponding entries of A and B have the same moduli. Hence $\Omega(A)$ is referred to as the *equimodular set of matrices* associated with A. It is helpful to observe that both A and M_A belong to $\Omega(A)$. Among other statements, Varga (Theorem 1, [132]) proves that for any $A \in \mathbb{C}^{n\times n}$ the comparison matrix M_A is of generalized positive type if and only if there exists $u \in \mathbb{R}^n$ such that $u > 0$ and $M_A u > 0$. This, in turn, is shown to be equivalent to the statement that every $B \in \Omega(A)$ is a nonsingular matrix. These are among a list of eleven equivalent statements which are equivalent to, or sharper than some of the extensions of diagonal dominance, obtained till then. The author notes that certain results in the literature are captured among these. More pertinent to the discussion at hand, one may now observe that the result of Shivakumar and Chew mentioned in the second paragraph of this section is a particular case of the result of Varga with $u = e$ (the vector all of whose coordinates are 1) and is weaker than the equivalence of the last two statements mentioned here.

Let us turn our attention to the chain condition stated in the beginning of this section. This condition (in addition to diagonal dominance and the nonemptiness assumption on J) is referred to as *weakly chained diagonal dominance* in [120]. We shall look at some pertinent results from this work. However, before doing this, let us paraphrase the result (Corollary 4, [103]) stated earlier as: A weakly chained diagonally dominant Z-matrix whose diagonals are positive is an invertible M-matrix. We now consider the work reported in [120]. It is pertinent to point to a couple of results, viz., one on upper bounds of the infinity norm of the inverse of a weakly chained diagonally dominant M-matrix A and another, on lower bounds for the entries of the inverse of A. Let B denote the principal submatrix obtained from A by deleting the first row and the first column. Then the authors show that B is also a weakly chained diagonally dominant M-matrix (Lemma 2.3, [120]).

Set

$$r_i = \frac{1}{|a_{ii}|} \sum\nolimits_{j=i+1}^{n} |a_{ij}|, \text{ with } r_n = 0.$$

Note that these are partial right row sums. We may now state one of the main results (Theorem 2.3, [120]). Let A be a weakly chained diagonally dominant M-matrix such that for all $k \in N$ one has $r_k < 1$. Then (the diagonal entries of A are positive and)

$$\|A^{-1}\|_\infty \leq \sum\nolimits_{i=1}^{n} \frac{1}{a_{ii} \prod_{j=1}^{i} (1 - r_j)}.$$

In order to state the next result, define

$$d_i = \frac{1}{|a_{ii}|}\sum_{j=i+1}^{n} |a_{ji}|, \text{ where we set } d_n = 0.$$

Note that these quantities are partial lower column sums. We have the following result: Let A be a weakly chained diagonally dominant M-matrix and $(f_{ij}) = F = A^{-1}$. Then (the diagonal entries of A are positive and) one has

$$\min_{j,k} f_{jk} \geq \frac{1}{a_{nn}}\prod_{i=1}^{n-1} \min\{d_i, r_i\}.$$

The authors also obtain upper and lower bounds for the least eigenvalue of A (the Perron root of A^{-1}) and also the corresponding normalized positive eigenvector (the Perron vector of A^{-1}).

We will later discuss an application of these results to a certain linear system of differential equations and digital circuit dynamics. This will appear in Sect. 8.2 of Chap. 8.

Let us include an improved result of the upper bound on the infinity norm of the inverse matrix, given as above. We observe that this result, to follow, is applicable to a subclass of the class of matrices considered above and was proved by Li [65]. Let A be a weakly chained diagonally dominant M-matrix. Set

$$s_n = \sum_{k=1}^{n-1} |a_{nk}|$$

and define recursively, for $k = n-1, n-2, \ldots, 1$

$$s_k = \sum_{i=1}^{k-1} |a_{ki}| + \sum_{i=k+1}^{n} |a_{ki}|\frac{s_i}{|a_{ii}|}.$$

Set $h_1 = 1$. Define $t_i = \sum_{k=1}^{n} |a_{ik}|$. Define recursively, for $k = 2, 3, \ldots, n$

$$h_k = t_{k-1} - s_{k-1}.$$

Suppose that for all $k \in N$, we have $|a_{kk}| + l_k > s_k$, where l_k are defined by the equation $|L|e = (l_1, l_2, \ldots, l_n)^T$. Here $(|l_{ij}|) = |L|$, where L is the strict lower triangular part of A (not including the diagonal). Li (Theorem 2.4, [65]) shows that

$$\|A^{-1}\|_\infty \leq \sum_{i=1}^{n} \prod_{k=1}^{i} \frac{h_k}{a_{kk} + l_k - s_k}.$$

Let us also report one more recent work, briefly. Huang and Zhu [49] obtained a new upper bound (for, once again a subclass of weakly chained diagonally dominant M-matrices) better than the result of Li mentioned above. The main result is Theorem 2, the details of which are too complicated to be included here. The authors include numerical examples to show the effectiveness of the new bound.

In the literature, many improvements to the various bounds discussed above have been presented. In this last part, we include one of the more recent results. Let A be a strictly row diagonally dominant M-matrix. Let the numbers r_i be defined as before for $i = 1, 2, \ldots, n-1$ with $r_n = 0$. Define

$$p_k = \max\{\frac{\sum_{j=1}^{n} |a_{ij}|}{|a_{ii}|} : k \leq i \leq n\}$$

with $p_n = 0$. For certain specific constants ω_{ij}, $i, j = 1, 2, \ldots, n$, Wang, Sun and Zhao (Theorem 2, [137]) have shown that

$$\|A^{-1}\|_\infty \leq \frac{1}{a_{11} - \sum_{k=2}^{n} |a_{1k}|\omega_{k1}} + \sum_{i=2}^{n} \{\frac{1}{a_{ii} - \sum_{k=i+1}^{n} |a_{ik}|\omega_{ki}} \prod_{j=1}^{i-1} \frac{1}{1 - r_j p_j}\}.$$

Numerical examples are used by these authors to illustrate that these are indeed improvements to earlier results.

As we reported earlier in this section, some recent results on H-matrices were obtained in the work [35]. Let us conclude this section with a passing remark that in Theorem 3 of that work, the authors prove a result using a chain condition on the entries of A ensuring that A is an H-matrix.

2.4 Tridiagonal Matrices

Tridiagonal matrices, finite or infinite, occur in a large number of applications. Some of these include solution methods by finite difference approximations for certain boundary value problems, cubic spline approximations, curve tracing, and three-term difference equations. Infinite tridiagonal matrices arise specifically in the solution of Mathieu's equation [119] and three-term recurrence relations for Bessel functions.

Here, A is a tridiagonal matrix denoted by $\{a_i, b_i, c_i\}$, where $b_i, i = 1, \ldots, n$ denote the diagonal elements, $a_i, i = 2 \ldots, n$ denote the lower diagonal elements and $c_i, i = 1, \ldots, n-1$ denote the upper diagonal elements. We assume that A is strictly row diagonally dominant with

$$\sigma_i |b_i| = |a_i| + |c_i|, \quad a_i, b_i, c_i \neq 0, \quad i = 1, 2, \ldots, n. \tag{2.8}$$

We then have

$$0 \leq \sigma_i < 1, \; i = 1, 2, \ldots, n.$$

A result of Ostrowski, as mentioned in Sect. 2.2, gives bounds on the inverse elements of a strictly row diagonally dominant matrix. It was also stated that some of these bounds could be improved for tridiagonal matrices. Shivakumar and Ji obtained one such specific result (Theorem 2.1, [108]), which we recall next. It is helpful to recall that the entries of the inverse of A are given by $\frac{A_{ji}}{\det A}$. Among other things, it was shown there that for the upper triangular entries (including the diagonal entries) ($i \le j$) one has

$$\left(\prod_{k=i+1}^{j} \frac{|a_k|}{|b_k|(1+\sigma_k)}\right) |A_{ii}| \le |A_{ij}| \le \left(\prod_{k=i+1}^{j} \sigma_k\right) |A_{ii}| \tag{2.9}$$

whereas for the lower triangular entries ($i > j$) the inequalities are given by

$$\left(\prod_{k=j}^{i-1} \frac{|c_k|}{|b_k|(1+\sigma_k)}\right) |A_{ii}| \le |A_{ij}| \le \left(\prod_{k=j}^{i-1} \sigma_k\right) |A_{ii}|. \tag{2.10}$$

One may verify now that these inequalities together improve upon the (lower and upper) bounds of Ostrowski, stated earlier. It is further shown (Theorem 2.2, [108]) that

$$\frac{1}{|b_i| + |a_i|\sigma_{i-1} + |c_i|\sigma_{i+1}} \le \left|\frac{A_{ii}}{\det A}\right| \le \frac{1}{|b_i| - |a_i|\sigma_{i-1} - |c_i|\sigma_{i+1}}, \tag{2.11}$$

where $\sigma_0 = \sigma_{n+1} = 0$.

The authors then establish the following inequalities for the inverse $F = (f_{ij})$ of A (Theorem 2.3, [108]):

$$\frac{\prod_{k=i+1}^{j} |a_k|}{\prod_{k=i}^{j} |b_k|(1+\sigma_k)} \le |f_{ji}| \le \frac{\prod_{k=i+1}^{j} \sigma_k}{|b_i| - |a_i|\sigma_{i-1} - |c_i|\sigma_{i+1}}, \; i \le j \tag{2.12}$$

giving bounds for the entries in the lower triangular part including the diagonals. For the upper triangular part one has

$$\frac{\prod_{k=j}^{i} -1|c_k|}{\prod_{k=j}^{i} |b_k|(1+\sigma_k)} \le |f_{ji}| \le \frac{\prod_{k=j}^{i-1} \sigma_k}{|b_i| - |a_i|\sigma_{i-1} - |c_i|\sigma_{i+1}}, \; i > j. \tag{2.13}$$

As mentioned earlier, upper bounds for $\|A^{-1}\|_\infty$ are relevant in connection with solutions of systems of linear equations. We present one more upper bound in what follows. Set

$$\sigma = \max_k\{\sigma_k\} \quad \text{and} \qquad \delta = \min_k\{|b_k|\}.$$

Using the results obtained above, an upper bound (as in the proof of Theorem 2.4) [108] is given by

$$\|A^{-1}\|_\infty \le \frac{1}{\delta(1-\sigma)}.$$

Note that the inequality fails if at least one σ_i equals one. The following extension is proved in that case. We quote the following (Theorem 2.5, [108]): Let A be a tridiagonal matrix defined as above with $\sigma_k < 1$ for some k and $\sigma_i = 1$ whenever $i \neq k$. Set

$$q = \min_l \left\{ \frac{|a_l|}{|b_l|}, \frac{|c_l|}{|b_l|} \right\},$$

$r = \frac{(1+q)}{q} > 2$ and $\delta = \min_k\{|b_k|\}$. Then

$$\|A^{-1}\|_\infty < \frac{2\left(r^{n+1} - r^{\frac{n+1}{2}}\right)}{\delta r^3}. \tag{2.14}$$

Next, we review some new upper and lower bounds for the entries of the inverses of diagonally dominant tridiagonal matrices obtained by Nabben [75], which improve the results of Shivakumar and Ji, described above. Let us state the main results, here. As before, assume that A is a tridiagonal matrix denoted by $\{a_i, b_i, c_i\}$, where $b_i, i = 1, \ldots, n$ denote the diagonal elements, $a_i, i = 2\ldots,n$ denote the lower diagonal elements, and $c_i, i = 1, \ldots, n-1$ denote the upper diagonal elements. Define

$$\tau_i = \frac{|c_i|}{|b_i - a_{i-1}|}, i = 1, 2, \ldots, n$$

and

$$\omega_i = \frac{|a_{i-1}|}{|b_i - c_i|}, i = 2, 3, \ldots, n,$$

where we tacitly assume that all the denominators are strictly positive. Note that if A is diagonally dominant, then one has $\tau_i \le 1$ and $\omega_i \le 1$ for each i. Again, let us denote $(f_{ij}) = F = A^{-1}$. One has the following: (Theorem 3.1, [75]) Let A be a nonsingular tridiagonal matrix given as above. If A is diagonally dominant, then (for the strictly upper triangular entries of F)

$$|f_{ij}| \leq |f_{jj}| \prod_{k=i}^{j-1} \tau_k, \; i < j$$

and (for the strictly lower triangular entries of F)

$$|f_{ij}| \leq |f_{jj}| \prod_{k=j+1}^{i} \omega_k, \; i > j.$$

Using this result, the following bounds are obtained (Theorem 3.2):

$$\frac{1}{|b_i| + \tau_{i-1}|a_{i-1}| + \omega_{i+1}|c_i|} \leq |f_{ii}| \leq \frac{1}{|b_i| - \tau_{i-1}|a_{i-1}| - \omega_{i+1}|c_i|},$$

if the denominator of the upper bound is nonzero. The fact that these improve upon the results stated above, among other things, follows from the discussion after Theorem 4.1, where a comparative analysis is presented.

Further, the author studies an iterative refinement for upper bounds. He points out that this iterative refinement gives the inverse of M_A (the comparison matrix of A) after $n-1$ iterations. If A is an M-matrix, then the iterative refinement produces the inverse of A. The author studies also the parallel implementation of the bounds and their computation for tridiagonal Toeplitz matrices.

Let us also mention in the passing that Nabben [76] proves structural characterizations for inverses of tridiagonal and banded matrices in the cases when A is an M-matrix, positive definite, or is diagonally dominant. In this work it is observed that certain entries along a row or column tend to decay in magnitude away from the diagonal, but for certain specific class of matrices they do not. The author investigates this phenomenon.

The results of Nabben above have been improved by Peluso and Politi [84]. We recall their main result (Theorem 4.1, [84]). Let A be a nonsingular tridiagonal matrix with the notation as above and $F = A^{-1}$. Then for certain specific constants δ_{kl} and γ_{kl} (which we do not define here) and for $l = 1, 2, \ldots, n-1$, one has

$$|f_{jj}| \prod_{k=i}^{j-1} \delta_{kl} \leq |f_{ij}|, \; i = 1, 2, \ldots, j-1$$

and

$$|f_{jj}| \prod_{k=j+1}^{i} \gamma_{kl} \leq |f_{ij}|, \; i = j+1, j+2, \ldots, n.$$

The authors include arguments to show how these are indeed improved bounds for all the entries of the inverse and in particular much better bounds for the diagonals of the inverse. Several numerical examples are presented to substantiate these statements.

2.5 Sign Patterns

It is observed that matrices resulting from many physical models have certain sign distributions. It remains both an important and an interesting problem to study how these distributions are related to nonsingularity.

We say that two matrices $A, B \in \mathbb{R}^{n \times n}$ have the same sign pattern if the signs of their corresponding entries are the same. A is called *sign nonsingular* if every matrix with the same sign pattern as A is nonsingular. This is obviously a much stronger form of nonsingularity. Define

$$\mathcal{S}_A := \{A \circ R : R > 0,\ R \in \mathbb{R}^{n \times n}\},$$

where $\circ$ denotes the Hadamard entrywise product and $R > 0$ denotes that all the entries of R are positive, as usual. It is now apparent that A is sign nonsingular if and only if every matrix in $\mathcal{S}_A$ is nonsingular. In what follows, we briefly review the work of Drew, Johnson, and van den Driessche on sign nonsingular matrices [25]. For $k = 1, 2, \ldots, n$, define

$$R_{n,k} := \{R \in \mathbb{R}^{n \times n} : R > 0,\ rank(R) \leq k\}$$

and

$$L_{n,k} := \{A \in \mathbb{R}^{n \times n} : R \in R_{n,k} \Longrightarrow \det(A \circ R) \neq 0\}.$$

Observe that $L_{n,n}$ is the set of all $(n \times n)$ sign nonsingular matrices and $L_{n,1}$ is the set of all (usual) nonsingular matrices. The authors introduce a second sequence of classes, which we recall below:

$$R'_{n,k} := \{R \in \mathbb{R}^{n \times n} : R > 0 \text{ and } R \text{ has at least } n - k + 1 \text{ rows of } 1\}$$

and

$$L'_{n,k} := \{A \in \mathbb{R}^{n \times n} : R \in R'_{n,k} \Longrightarrow \det(A \circ R) \neq 0\}.$$

Among other things, it is shown that for $n = 2, 3, 4$ each class of the sequence $L_{n,k}$ is distinct (Theorem 4.4), $L_{5,4} = L_{5,5}$ (Theorem 4.5) and $L_{n,2} = L_{n,n}$, for $n \geq 15$ (Theorem 4.8). Let us also include another interesting result (Theorem 4.9) that was proved: $A \in L'_{n,k}$ if and only if A is nonsingular and the product $A \circ A^{-T}$ is a doubly stochastic matrix. A^{-T} denotes inverse transposed.

Next, we turn our attention to two recently obtained nonsingularity criteria based on sign distributions and some stated conditions on the elements of A [109]. These considerations were motivated by a study of Poisson's equation for doubly connected regions.

Let us suppose that the sign distribution for the matrix A is given as one of the following [109]: Let us refer to the first class as matrices of type I. These matrices are defined by $a_{ij} > 0$ for $i \leq j$ and

$$(-1)^{i+j} a_{ij} > 0 \text{ for } i > j, \quad i,j = 1,2,\ldots,n. \tag{2.15}$$

Type II matrices are those that satisfy $a_{ij} > 0$ for $i \leq j$ and

$$(-1)^{i+j+1} a_{ij} > 0 \text{ for } i > j, \quad i,j = 1,2,\ldots,n. \tag{2.16}$$

Let us describe these matrices in words. First, observe that all the entries are nonzero. Matrices of both types have the property that all the entries in the upper triangular part including the diagonal are positive. To describe the lower triangular entries, let us note that matrices of the first type satisfy the condition that the entries in the first column alternate in sign (with the first entry being already positive), the entries in the second column alternate in sign from the second diagonal entry onwards (which is already positive, again), the entries in the third column alternate in sign from the third (positive) diagonal element onwards, and so on. The lower triangular entries for matrices of the second type follow an almost similar pattern as the first type, but with a shift. Namely, the entries in the first column alternate in sign, this time from the second row, entries in the second column alternate in sign from the third row, and so on.

In order to state the result on nonsingularity, for convenience we define

$$\nu_{ij} = a_{ij} - \sum_{k=j+1}^{m} a_{ik}, \; i \leq j \tag{2.17}$$

and

$$\mu_{ij} = a_{ij} - \sum_{k=1}^{i-1} a_{kj}, \; i \leq j \tag{2.18}$$

and for $j < i < l < n$,

$$\omega_{ij}^{l} = \min\left\{ |a_{ij}| - \left| \sum_{k=i+1}^{l} a_{kj} \right|, \left| \sum_{k=i+1}^{l} a_{kj} \right| - |a_{l+1,j}| \right\} \tag{2.19}$$

together with

$$\delta_{ij}^{l} = \left| \sum_{k=j+1}^{l} a_{ik} \right| - |a_{ij}|, \text{ for } j < l \leq i. \tag{2.20}$$

Now we state the main result (Theorem 2.1) of Shivakumar and Ji [109]. Let A be a real matrix of one of the two types described as above. Suppose that one has

$$\nu_{ij} > 0, \tag{2.21}$$

$$\mu_{ij} > \sum_{k=j+1}^{n} \mu_{ik} > 0, \tag{2.22}$$

for $i \leq j$ and

$$\omega_{ij}^{l} > 0, \ (j < i < l), \tag{2.23}$$

$$\delta_{ij}^{l} > \sum_{k=i+1}^{n} \delta_{kj}^{l} > 0, (j < l \leq i), \tag{2.24}$$

for $i > j$. Then A is nonsingular.

Informally, the positivity of ν_{ij} and ω_{ij} shows a decreasing absolute value for the upper triangular entries of A along the horizontal direction and of the lower triangular entries along the vertical downward direction, respectively. The other inequalities may be interpreted as second order distribution properties, reflecting a similar trend along the horizontal and the vertical directions, as above. Finally, one may view all these inequalities as describing a certain scattering distribution along the horizontal and the vertical directions for any element of A. Observe that these conditions on a matrix can be verified easily and efficiently. Let us just mention in the passing that the proof uses induction on the order of the matrix involving some very complicated calculations!

Let us conclude this section with a numerical example. Let $A = (a_{ij})$ be the matrix whose upper triangular entries (including the diagonal entries) are given by $\frac{1}{2^{j-i}}$ and whose lower triangular entries are $(-\frac{1}{2})^{i-j}$. This is a matrix of sign distribution of the first type and also satisfying the hypotheses of the result given above. Hence it is nonsingular. Note, however, that it is not diagonally dominant.

Chapter 3
Infinite Linear Equations

3.1 Introduction

In this chapter, we shall first review, in Sect. 3.2, certain results on infinite linear systems whose "coefficient" matrices are diagonally dominant in some sense. We treat these infinite matrices in their own right and also as operators over certain normed linear spaces. These results show the extent to which results that are known for finite matrices have been generalized. Next, in Sect. 3.3, we recall some results on eigenvalues for operators mainly of the type considered in the second section. We also review a powerful numerical method for computing eigenvalues of certain diagonally dominant tridiagonal operators. Section 3.4 concerns linear differential systems whose coefficient matrices are operators on either ℓ^1 or ℓ^∞. Convergence results for truncated systems are presented. The concluding section, viz., Sect. 3.5 discusses an iterative method for numerically solving a linear equation whose matrix is treated as an operator on ℓ^∞ satisfying certain conditions, including a diagonal dominance condition.

3.2 Infinite Linear Systems

In this section, we shall take a look at considerations of diagonal dominance for infinite matrices. In classical analysis, linear equations in infinite matrices occur in problems including interpolation, sequence spaces, and summability theory. An earlier notable result (considered till then) on the existence of a solution to such an infinite system was given by Polya, which, however, excluded discussion of uniqueness. Kantorovich and Krylov [54] state certain results (without proofs) which provide sufficient conditions for the existence and uniqueness of bounded

P.N. Shivakumar et al., *Infinite Matrices and Their Recent Applications*,
DOI 10.1007/978-3-319-30180-8_3

solutions under the assumption that the infinite matrix under consideration is invertible. One of the motivations for the results presented in this section is in solving certain elliptic partial differential equations in multiply connected regions.

In what follows, we shall first review two works, viz., [113] and [101], where infinite matrices were considered not as operators on some normed linear spaces. Specifically, we are concerned here with the infinite system of linear equations [113]:

$$\sum_{j=1}^{\infty} a_{ij}x_j = b_i, \; i \in \mathbb{N}, \tag{3.1}$$

or alternatively

$$Ax = b, \tag{3.2}$$

where the infinite matrix $A = (a_{ij})$, is strictly "column" diagonally dominant, i.e., there exist numbers σ_i with $0 \leq \sigma_i < 1$, $i \in \mathbb{N}$ such that

$$\sigma_i |a_{ii}| = \sum_{j=1}^{\infty} |a_{ij}|, \tag{3.3}$$

where, of course, $a_{ii} \neq 0$ for all $i \in \mathbb{N}$ and the sequence $\{b_i\}$ is bounded. One is interested in sufficient conditions that guarantee the existence and uniqueness of a bounded solution to the above system. The idea of the approach taken in [113] is to use finite truncations and develop estimates for such a truncated system, viz.,

$$\sum_{j=1}^{n} a_{ij}x_j = b_i, \; i = 1, 2, \ldots, n. \tag{3.4}$$

Using $A^{(n)}$ to denote the matrix obtained from A by taking the first n rows and n columns it may be shown that $\det(A^{(n)}) \neq 0$ for each n. Adopting a similar notation for $x^{(n)}$ and $b^{(n)}$, the truncated system above could be rewritten as

$$A^{(n)}x^{(n)} = b^{(n)}. \tag{3.5}$$

Let us denote the unique solution of this truncated solution by $x^{(n)}$. The following inequalities are established: for each $j \geq 1$ and $n \geq j$, one has

$$\left| x_j^{(n+1)} - x_j^{(n)} \right| \leq P\sigma_{n+1} + \frac{Q}{|a_{n+1,n+1}|} \tag{3.6}$$

for some positive constants P and Q. For any two positive integers p, q and for each fixed j, $j \leq p, q$ one also has

$$\left|x_j^{(q)} - x_j^{(p)}\right| \leq P \sum_{i=p+1}^{\infty} \sigma_i + Q \sum_{i=p+1}^{q} \frac{1}{|a_{i,i}|}. \tag{3.7}$$

Using standard estimates for strictly row diagonally dominant finite systems, an estimate for the solution of the truncated system is given by,

$$\left|x_j^{(n)}\right| \leq \prod_{k=1}^{n} \frac{1+\sigma_k}{1-\sigma_k} \sum_{k=1}^{n} \frac{|b_k|}{|a_{kk}|(1+\sigma_k)}, \tag{3.8}$$

for each j with $j \leq n$. Turning the attention to the infinite system, let us assume that one has the following for the entries a_{ij}:

$$\sum_{i=1}^{\infty} \frac{1}{|a_{ii}|} < \infty, \tag{3.9}$$

and for some $M > 0$ and all $i \in \mathbb{N}$

$$\sum_{j=1, j\neq i}^{\infty} |a_{ij}| \leq M. \tag{3.10}$$

Then Shivakumar and Wong show (Theorems 1 and 2, [113]) that the infinite system considered above has a unique and a bounded solution. Let us observe that the authors give a numerical example to show that a general infinite system which has a unique bounded solution could still have unbounded solutions.

Shivakumar (Theorems 3 and 4, [101]), later relaxed the assumption on the absolutely summability of the reciprocals of the diagonals of A, while retaining the other assumptions to show that one could still recover similar results, like existence and uniqueness. In the presence of another rather strong assumption, he shows that A^{-1} is also strictly "row" diagonally dominant. It must be remarked that this is a rather unusual result, especially for infinite matrices.

Now, we take the point of view of studying infinite matrices as operators over certain Banach spaces. We will be discussing two specific instances of bounded operators over Banach spaces. These are the spaces: ℓ^1, the space of absolutely summable complex sequences and ℓ^∞, the space of bounded complex sequences. We shall review some recent results on certain classes of strictly ("row" or "column") diagonally dominant infinite matrices that turn out to be invertible. Bounds on the inverses in these cases are given. The work reported here is due to Shivakumar, Williams and Rudraiah [119].

Given a matrix $A = (a_{ij}),\ i,j \in \mathbb{N}$, a space of infinite sequences X over the real or the complex field and $x = (x_i),\ i \in \mathbb{N}$, we define Ax by

$$(Ax)_i = \sum_{j=1}^{\infty} a_{ij}x_j,$$

provided this series converges for each $i \in \mathbb{N}$. We define the domain of A by

$$D(A) = \{x \in X : Ax \text{ exists and } Ax \in X\}.$$

Let us start with the case $X = \ell^1$. Consider an infinite matrix A on ℓ^1. We assume the following: Suppose that the "diagonals" of A are all nonzero and form an unbounded sequence of real or complex numbers. Let A be uniformly strictly "column" diagonally dominant in the sense that the following condition holds: There exist numbers $\rho,\ 0 \le \rho < 1$ and $\rho_j,\ 0 \le \rho_j \le \rho$ such that one has

$$Q_j = \sum\nolimits_{i=1, i\neq j}^{\infty} |a_{ij}| = \rho_j |a_{jj}|,\ j \in \mathbb{N}.$$

We further assume that

$$|a_{ii} - a_{jj}| \ge Q_i + Q_j, \text{ for all } i,j \in \mathbb{N},\ i \neq j$$

and

$$\sup\{|a_{ij}| : j \in \mathbb{N}\} < \infty \text{ for all } i \in \mathbb{N}.$$

For an operator A satisfying the first two conditions, Shivakumar, Williams, and Rudraiah show (Theorem 2, [119]) that A is an operator with dense domain, is invertible, and A^{-1} is compact. The following upper bound is also proved:

$$\|A^{-1}\|_1 \le \frac{1}{(1-\rho)(\inf_i |a_{ii}|)}.$$

Similar results are also derived for operators on ℓ^∞. For an operator A on ℓ^∞ consider the following set of conditions which could be considered "dual" to the assumptions that were made for an operator on ℓ^1 listed above. There exist numbers $\sigma,\ 0 \le \sigma < 1$ and $\sigma_j,\ 0 \le \sigma_j \le \sigma$ such that one has

$$P_i = \sum\nolimits_{j=1, j\neq i}^{\infty} |a_{ij}| = \sigma_j |a_{jj}|,\ i \in \mathbb{N},$$

$$|a_{ii} - a_{jj}| \ge P_i + P_j, \text{ for all } i,j \in \mathbb{N},\ i \neq j$$

and

$$\sup\{|a_{ij}| : i \in \mathbb{N}\} < \infty \text{ for all } j \in \mathbb{N}.$$

Analogous to the case of ℓ^1, the first condition could be considered as a uniform strict "row" diagonal dominance. Then, for an operator A on ℓ^∞ satisfying the two conditions given above, along with the condition that the main diagonal elements of A are all nonzero and form an unbounded sequence, it is proved (Theorem 4, [119]) that A is a closed operator, A^{-1} exists, A^{-1} is compact, and

$$\|A^{-1}\|_\infty \leq \frac{1}{(1-\sigma)(\inf_i |a_{ii}|)}.$$

It is clear that this last inequality is a generalization of the inequality of Varah mentioned in Sect. 2.2, for infinite matrices. It must be remarked here that there is a departure from what one experiences in the finite matrix case for a diagonally dominant matrix which is also irreducible. The authors of the work reported here demonstrate by an example (Example 1, [119]) that there are infinite matrices (considered as bounded operators on ℓ^∞) which are irreducible and diagonally dominant (meaning that one has $\sigma = 1$ with all $\sigma_i < 1$) but are not invertible.

We close this section by mentioning some recent results on infinite matrices that were obtained by Williams and Ye [138], which, however, do not concern either diagonal dominance or invertibility. This work investigates conditions that guarantee when an infinite matrix will be bounded as an operator on two weighted ℓ^1 spaces and obtains a relationship between such a matrix and the given weight vector. It is established that every infinite matrix is bounded as an operator between two weighted ℓ^1 spaces for suitable weights. Necessary conditions and separate sufficient conditions for an infinite matrix to be bounded on a weighted ℓ^1 space, with the same weight for the domain and codomain, are presented.

3.3 Linear Eigenvalue Problem

In this section, we consider the eigenvalue problem for infinite matrices considered as operators on certain Banach spaces. We also discuss results on the problem of determining the location of eigenvalues for diagonally dominant infinite matrices and determining upper and lower bounds for them. First, we report the results of Shivakumar, Williams and Rudraiah [119].

If $A = (a_{ij})$, $i, j \in \mathbb{N}$, and X is a space of infinite sequences, then the domain of A denoted by $D(A)$ is defined as in the previous section. We define an eigenvalue of A to be any scalar λ (from the underlying field) for which $Ax = \lambda x$ for some $0 \neq x \in D(A)$. We define the Gershgorin disks by considering A as an operator on ℓ^1, by

$$C_i = \{z \in \mathbb{C} : |z - a_{ii}| \leq Q_i\},\ i \in \mathbb{N},$$

where the numbers Q_i are as defined in the previous section. Let us also recall another assumption that was made there:

$$|a_{ii} - a_{jj}| \geq Q_i + Q_j, \text{ for all } i, j \in \mathbb{N}, \ i \neq j.$$

We observe that this condition on A is equivalent to the (almost) disjointness of the Gershgorin disks C_i, viz., the intersection of two disks consists of at most one point. Finally, the condition on the boundedness of the suprema made as in the earlier section implies that A is a closed linear operator on ℓ^1.

One of the main results of the work being reported here is in showing the following for an operator A on ℓ^1 satisfying the three conditions of Sect. 3.2. This result (Theorem 3, [119]) states that A consists of a discrete countable set of nonzero eigenvalues $\{\lambda_k : k \in \mathbb{N}\}$ such that $|\lambda_k| \to \infty$ as $k \to \infty$.

For the case of ℓ^∞, we assume that the entries a_{ij} satisfy those conditions that are given in Sect. 3.2. We define the Gershgorin disks as

$$D_i = \{z \in \mathbb{C} : |z - a_{ii}| \leq P_i\}, \ i \in \mathbb{N}.$$

The authors also prove another result (Theorem 5, [119]) similar to the ℓ^1 case, that the spectrum of A (satisfying all the three conditions listed above) consists of a discrete countable set of nonzero eigenvalues $\{\lambda_k : k \in \mathbb{N}\}$ such that $|\lambda_k| \to \infty$ as $k \to \infty$.

Let us mention certain interesting contributions and generalizations of the work reported earlier in this section, that have been made by Farid and Lancaster [32] and [33]. In the first work, certain Gerschgorin type theorems were established for a class of row diagonally dominant infinite matrices by considering them as operators on ℓ^p spaces, $1 \leq p \leq \infty$. The authors develop a theory analogous to the work in [119]. They provide constructive proofs where a sequence of matrix operators is shown to converge (in the sense of the gap for closed operators) to the diagonally dominant operator that one started with. Utilizing eigenvalues and eigenvectors of such a sequence of matrix operators, the problem of convergence of these eigenvalues and the corresponding eigenvectors to a simple eigenvalue and the corresponding eigenvector of the given operator, is investigated. Despite the fact that the range of the value p was extended in this work, the results here for $p = 1$ and $p = \infty$ are weaker than the corresponding ones of [119]. However, in the second work, the authors show how the earlier contributions of [119] could be both strengthened and extended to more general values of p. Here, row diagonally dominant infinite matrices are considered as closed operators with compact inverses on ℓ^p spaces, $1 \leq p \leq \infty$. The authors extend the results of their earlier work for the case of $p = 1$ and $p = \infty$. Results for column diagonally dominant infinite matrices are also derived (Theorems 2.1, 3.1 and 3.2, [33]).

Let us include some other contributions, as well. Farid [29] shows (Theorem 3.2) that the eigenvalues of a diagonally dominant infinite matrix satisfying certain additional conditions, acting as a linear operator in ℓ^2 approach its main diagonal. He also discusses an application of this result to approximate the eigenvalues of the Mathieu's equation. Malejki [70] studies a real symmetric tridiagonal matrix A whose diagonal entries and off-diagonal entries satisfy certain decay properties. It follows that such an operator A has a discrete spectrum. Let $A^{(n)}$ be its $n \times n$ truncation. The main result of the author is in showing the following: If the eigenvalues of A are

$$\lambda_1 \leq \lambda_2 \leq \ldots$$

and

$$\mu_{1,n} \leq \mu_{2,n} \leq \ldots \mu_{n,n}$$

are the eigenvalues of $A^{(n)}$, then for every $\gamma > 0$ and $r \in (0, 1)$, there exists a constant c such that

$$|\mu_{k,n} - \lambda_k| \leq cn^{-\gamma} \text{ for all } 1 \leq k \leq rn.$$

To conclude this section, we discuss a powerful computational technique for determining the eigenvalues of the infinite system $Ax = \lambda x$ derived by using a truncated matrix $G^{(1,k)}$, to be defined below. The idea of this technique is to box the eigenvalues and then use a simple bisection method to give the value of λ_n to any required degree of accuracy.

Consider a matrix $A = (a_{ij})$ acting on ℓ^∞ satisfying all the four conditions given earlier and satisfy in addition the following:

$$a_{ij} = 0, \text{ if } |i-j| \geq 2, \ i, j \in \mathbb{N}$$

$$0 < a_{ii} < a_{i+1,i+1}, \ i \in \mathbb{N}$$

and

$$a_{i,i+1}a_{i+1,i} > 0, \ i \in \mathbb{N}.$$

Observe that the first condition here means that A is a tridiagonal matrix, viz., the entries not in the principal diagonal and the two immediate subdiagonals (the lower and the upper) are zero.

Suppose that the scalar λ satisfies

$$a_{nn} - P_n \leq \lambda \leq a_{nn} + P_n, \text{ for all } n \in \mathbb{N},$$

where P_i are as defined above (in connection with the Gerschgorin circles). Let $G = A - \lambda I$. Let $G^{(1,k)}$ denote the truncated matrix of A obtained from A by taking only the first k rows and k columns. Denote $\beta_{1,k} := \det G^{(1,k)}$. We then have the following [119, Sect. 8]:

$$\beta_{1,k} = (a_{11} - \lambda)\beta_{2,k} - a_{12}a_{21}\beta_{3,k} \tag{3.11}$$

$$= [(a_{11} - \lambda)(a_{22} - \lambda) - a_{12}a_{21}]\beta_{3,k} - (a_{11} - \lambda)a_{23}a_{32}\beta_{4,k} \tag{3.12}$$

so that

$$\beta_{1,k} = p_s \beta_{s,k} - p_{s-1} a_{s-1,s} a_{s,s-1} \beta_{s+1,k}, \tag{3.13}$$

where the sequence p_s is defined by $p_0 = 0,\ p_1 = 1$ and

$$p_s = p_{s-1}\,(a_{s-1,s-1} - \lambda) - p_{s-2}\,a_{s-1,s-2}\,a_{s-2,s-1}.$$

If we set

$$Q_{s,k} = \frac{p_s}{p_{s-1}} - a_{s-1,s} a_{s,s-1} \frac{\beta_{s+1,k}}{\beta_{s,k}}, \tag{3.14}$$

we then have

$$\beta_{1,k} = p_{s-1} Q_{s,k} \beta_{s,k}. \tag{3.15}$$

We have the following cases:

Case (i): p_{s-1} and p_s have opposite signs.
Then $Q_{s,k} < 0$ and $\beta_{1,k}$ has the same sign as $-p_{s-1}$.

Case (ii): p_{s-1} and p_s have the same sign and

$$\frac{p_s}{p_{s-1}} > \frac{a_{s,s-1} a_{s-1,s}}{a_{ss} - \lambda - |a_{s,s+1}|}.$$

We then have $Q_{s,k} > 0$ and $\beta_{1,k}$ has the same sign as p_{s-1}.

Case (iii): p_{s-1} and p_s have the same sign and

$$\frac{p_s}{p_{s-1}} < \frac{a_{s,s-1} a_{s-1,s}}{a_{ss} - \lambda - |a_{s,s+1}|}.$$

Then, $Q_{s,k} < 0$ and $\beta_{1,k}$ have the same sign as $-p_{s-1}$.

We can use the method of bisection to establish both upper and lower bounds for λ_n to any degree of accuracy.

Let us close this section by mentioning in the passing that in this work, the authors (in [119], Sect. 6) establish some results concerning the convergence of the sequence of solutions of the truncated systems and study error analysis in detail. Application of the above technique to study Bessel functions forms the discussion in Sect. 8.6 of Chap. 8.

3.4 Linear Differential Systems

We consider the infinite linear system of differential equations:

$$\frac{d}{dt}x_i(t) = \sum_{j=1}^{\infty} a_{ij}x_j(t) + f_i(t), \qquad t \geq 0,\ x_i(0) = y_i,\ i \in \mathbb{N}, \tag{3.16}$$

where the functions f_i and numbers y_i are known. Using the notation $\mathbf{x}(t) = (x_i(t))$, $\mathbf{y} = (y_i)$ and $\mathbf{f}(t) = (f_i(t))$, this equation can be rewritten as

$$\dot{\mathbf{x}}(t) = A\mathbf{x}(t) + \mathbf{f}(t), \qquad \dot{\mathbf{x}}(0) = \mathbf{y}. \tag{3.17}$$

This equation is of considerable theoretical and applied interest. In particular, such systems occur frequently in topics including the theory of stochastic processes, perturbation theory of quantum mechanics, degradation of polynomials, and infinite ladder network theory. Arley and Brochsenius [3], Bellman [7], and Shaw [100] have made some notable contributions to the problem posed above. In particular, if A is a bounded operator on ℓ^1, then convergence of a truncated system has been established. However, none of these works yields explicit error bounds for such a truncation. In what follows, we recall the results of Shivakumar, Chew and Williams [118] for such error bounds, among other things. The analysis in this work concerns A, being a constant infinite matrix defining a bounded operator on X, where X is one of the spaces ℓ^1, ℓ^∞, or c_0, the latter being the space of complex sequences converging to zero. Explicit error bounds are obtained for the approximation of the solution of the infinite system by the solutions of finite truncation systems.

To begin with, we present the following framework for homogeneous systems ($\mathbf{f} = 0$): First, we assume that $\mathbf{y} \in \ell^1$. Next, suppose that

$$\alpha = \sup\{\sum_{i=1}^{\infty} |a_{ij}| : j \in \mathbb{N}\} < \infty.$$

Set

$$\gamma_n = \sup\{\sum_{i=n+1}^{\infty} |a_{ij}| : j = 1, 2, \ldots, n\}$$

and

$$\delta_n = \sup\{\sum_{i=1}^{n} |a_{ij}| : j = 1, 2, \ldots, n\}.$$

We assume that

$$\gamma_n \to 0 \text{ and } \delta_n \to 0, \text{ as } n \to \infty.$$

In the above, the finiteness of the supremum is equivalent to the statement that A is bounded on ℓ^1. The assumption involving γ_n states that the sums in the definition involving α converge uniformly below the main diagonal; it is a condition involving only the entries of A below the main diagonal. On the other hand, the assumption involving δ_n is a condition involving only the entries of A above the main diagonal.

For the sake of convenience and ease of use, let us adopt a notation used earlier for denoting a different object. So, let us define the matrix $A^{(n)}$ by: $(A^{(n)})_{ij} = a_{ij}$ if $1 \leq i, j \leq n$ and $(A^{(n)})_{ij} = 0$, otherwise. One applies similar definitions for $\mathbf{y}^{(n)}$ and $\mathbf{f}^{(n)}$. This leads to the definition

$$b_{ij}{}^{(n)} = ((A^{(n)})^j \mathbf{y}^{(n)})_i$$

using which we finally set

$$x_i^{(n)}(t) = \sum\nolimits_{j=1}^{\infty} \frac{t^j}{j!} b_{ij}{}^{(n)}, \ 1 \leq i \leq n.$$

In (Theorem 1, [118]) the following result is established: Suppose that the first two assumptions on A as given above are satisfied together with one of the next two conditions. Then

$$\lim\nolimits_{n \to \infty} \mathbf{x}^{(n)}(t) = \mathbf{x}(t)$$

in the l^1 norm uniformly in t on compact subsets of $[0, \infty)$. One also has explicit error bounds as given below:

$$\sum_{i=1}^{n} |x_i(t) - x_i^{(n)}(t)| \leq \alpha t e^{\alpha t} \left[\frac{1}{2} \gamma_n M t + \sum_{k=n+1}^{\infty} |\mathbf{y}_k| \right] \tag{3.18}$$

and

$$\sum_{i=n+1}^{\infty} |x_i(t)| \leq e^{\alpha t} \left[\gamma_n M t + \sum_{k=n+1}^{\infty} |\mathbf{y}_k| \right]. \tag{3.19}$$

Combining these, one has

$$\|\mathbf{x}(t) - \mathbf{x}^{(n)}(t)\| \leq e^{\alpha t} \left[\left(1 + \frac{1}{2} \alpha t\right) \gamma_n M t + (1 + \alpha t) \sum_{k=n+1}^{\infty} |\mathbf{y}_k| \right],$$

corresponding to the condition on γ_n (with the right-hand side converging to zero as $n \to \infty$) and

$$\sum\nolimits_{i=1}^{n} |x_i(t) - x_i^{(n)}(t)| \leq \delta_n M t e^{\alpha t},$$

corresponding to the condition on δ_n (with the right-hand side converging to zero as $n \to \infty$).

For the nonhomogeneous system, one assumes that f_i is continuous on $[0, \infty)$, $i \in \mathbb{N}$ and $\|\mathbf{f}(t)\| = \sum_{i=1}^{\infty} |f_i(t)|$ converges uniformly in t on compact subsets of $[0, \infty)$. If one defines $L(t) = \sup\{\|f(\tau)\| : 0 \le \tau \le t\}$, it then follows that the condition on $\mathbf{f}$ above is equivalent to the statement that $\mathbf{f}$ is continuous from $[0, \infty)$ into ℓ^1 and so one has $L(t) < \infty$ for all $t \ge 0$.

We also have the following result (Theorem 2, [118]): Suppose that one has the assumptions $\alpha < \infty$ and the condition on f_i given as earlier. In addition, suppose that either of the conditions on γ_n or δ_n hold. Then

$$\lim_{n\to\infty} \mathbf{x}^{(n)}(t) = \mathbf{x}(t)$$

in the l^1 norm uniformly in t on compact subsets of $[0, \infty)$, with explicit error bounds as given below:

$$\|\mathbf{x}(t) - \mathbf{x}^{(n)}(t)\| \le \frac{1}{2} t^2 e^{\alpha t} \gamma_n L(t) + t e^{\alpha} t \sup\{ \sum_{k=n+1}^{\infty} |f_k(\tau)| : 0 \le \tau \le t\},$$

corresponding to the condition on γ_n (with the right-hand side converging to zero as $n \to \infty$) and

$$\sum_{i=1}^{n} |x_i(t) - x_i^{(n)}(t)| \le \delta_n L(t) \alpha^{-2} \left[\alpha t e^{\alpha t} + (1 - e^{\alpha} t)\right], \tag{3.20}$$

corresponding to the condition on δ_n (with the right hand side converging to zero as $n \to \infty$).

Similar results hold for systems on l^{∞} (Theorems 4 and 6, [118]) and for systems on c_0 (Theorems 3 and 5, [118]). We refer the reader to [118] for details.

3.5 An Iterative Method

Iterative methods for linear equations in finite matrices have been the subject of a very vast literature. Since all these methods involve the nonsingularity of the matrix, the various notions of diagonal dominance of matrices have played a major role, as evidenced in our discussion in Chap. 2.2. The interest in this section is to discuss an iterative method for certain diagonally dominant infinite systems which we believe to be one of the first attempts towards such extensions.

Let us recall that in Sect. 2.3, we reviewed the work of Shivakumar and Chew where a certain notion of weakly chained diagonal dominance was discussed. A convergent iterative procedure was also proposed in the work [104].

In this section, we consider an infinite system of equations of the form $T\mathbf{x} = \mathbf{v}$, where $\mathbf{x}, \mathbf{v} \in \ell^\infty$, and T is a (possibly unbounded) linear operator on ℓ^∞. Suppose that the matrix of T relative to the usual Schauder basis is given by $T = (t_{ij})$. Consider the following:

1. There exists $\eta > 0$ such that $|t_{ii}| \geq \eta$ for all $i \in \mathbb{N}$.
2. There exist σ with $0 \leq \sigma < 1$ and σ_i, $0 \leq \sigma_i < \sigma < 1$, $i \in \mathbb{N}$ such that

$$\sum_{j=1,\, j\neq i}^{\infty} |t_{ij}| = \sigma_i |t_{ii}|,$$

3. $\sum_{j=1}^{i-1} \frac{|t_{ij}|}{|t_{ii}|} \to 0$ as $i \to \infty$.
4. Suppose further that either the diagonals of T form an unbounded sequence or that $\mathbf{v} \in c_0$.

Shivakumar and Williams first prove the following result (Theorem 1, [112]): Let $\mathbf{v} \in \ell^\infty$ and let T satisfy the first two conditions listed above. Then T has a (bounded) inverse and the equation $T\mathbf{x} = \mathbf{v}$ has a unique ℓ^∞ solution. This solution $\mathbf{x}$ satisfies the inequality (all the norms $\| \,.\, \|$ denote $\| \,.\, \|_\infty$):

$$\| \mathbf{x} \| = \| T^{-1}\mathbf{v} \| \leq \frac{\| \mathbf{v} \|}{\eta(1-\sigma)}.$$

It must be remembered that two results of a similar type from the work of [113] were discussed in Sect. 3.2.

Let $T = D + F$, where D is the main diagonal of T, (which, by virtue of the first assumption, is invertible) and F is the off-diagonal of T. Let A be defined by $A = -D^{-1}F$ and $\mathbf{b} = D^{-1}\mathbf{v}$. Then $T\mathbf{x} = \mathbf{v}$ is equivalent to the fixed-point system $\mathbf{x} = A\mathbf{x} + \mathbf{b}$, $\mathbf{b} \in \ell^\infty$, where A is a bounded linear operator on ℓ^∞. If one considers all the four conditions on A listed above, then one has the following consequences:

1. $\|A\| = \sup_{i\geq 1} \sum_{j=1}^{\infty} |a_{ij}| \leq \sigma < 1$,
2. $\sum_{j=1}^{i-1} |a_{ij}| \to 0$ as $i \to, \infty$ and
3. $\mathbf{b} = (b_i) \in c_0$.

Let us note that the fixed-point equation above leads naturally into the iterative scheme:

$$\mathbf{x}^{(p+1)} = A\mathbf{x}^{(p)} + \mathbf{b},\ \mathbf{x}(0) = \mathbf{b},\ p = 0, 1, 2, \ldots \tag{3.21}$$

As before, let $A^{(n)} = (a_{ij}^{(n)})$ be the infinite matrix obtained from A, where

$$a_{ij}^{(n)} = \begin{cases} a_{ij} & \text{if } 1 \leq i, j \leq n, \\ 0 & \text{otherwise.} \end{cases} \quad (3.22)$$

Thus $A^{(n)}$ is the $n \times n$ truncation of A padded with zeros. Let $\mathbf{x}^{(p,n)}$ be such that

$$(\mathbf{x}^{(p,n)})_i = b_i \text{ whenever } i > n.$$

By starting with $\mathbf{x}^{(0,n)} = \mathbf{b}$, we consider the following truncated iterations:

$$\mathbf{x}^{(p+1,n)} = A^{(n)}\mathbf{x}^{(p,n)} + b, \quad \mathbf{x}^{(0,n)} = \mathbf{b} \quad (3.23)$$

for $p = 0, 1, 2, \ldots$ and $n = 1, 2, 3 \ldots$. Then one has (Theorem 2, [112]): for certain constants β, β_n, γ_n, and μ_n

$$\|\mathbf{x}^{(p)} - \mathbf{x}^{(p,n)}\| \leq \beta\gamma_n \sum_{k=0}^{p-1}(k+1)\sigma^k + \beta_n\mu_n \sum_{k=0}^{p-1}\sigma^k, \quad (3.24)$$

where the right-hand side converges to zero as $n \to \infty$. It can also be shown that the following result holds:

Corollary 3.5.1 (Corollary 2, [112]).

$$\|\mathbf{x} - \mathbf{x}^{(p,n)}\| \leq \sigma^{p+1}(1-\sigma)^{-1}\beta + \beta\gamma_n(1-\sigma)^{-2} + \beta_n\mu_n(1-\sigma)^{-1}, \quad (3.25)$$

where the right-hand side converges to zero as $n \to \infty$.

An application of the above in the recurrence relations of the Bessel functions is given in Sect. 8.6 in Chap. 8.

Chapter 4
Generalized Inverses: Real or Complex Field

4.1 Introduction

The main objective of this chapter is to review certain recent results that were obtained in the context of generalized inverses of infinite matrices. These are presented in Sect. 4.2. We take this opportunity to review the basic ideas in the theory of generalized inverses of matrices and also operators acting between Hilbert spaces. This will be presented in the next section. We do not attempt at being exhaustive in our presentation. The intention is to give a brief idea of the notion of generalized inverses. Several excellent texts have been written on this topic. For matrices, for instance, we refer to the books by Ben-Israel and Greville [10], Meyer [73], and the classic text by Rao and Mitra [95]. For operators on general infinite dimensional spaces, see the book by Groetsch [41] and Nashed [77], where the latter includes extensive discussions on many algebraic as well as topological spaces. Let us note that in the next chapter, certain very new results on generalized inverses of matrices over quaternion polynomial rings are presented. For generalized inverses of matrices over commutative rings, an excellent source is the book by Bhaskara Rao [13].

For $A \in \mathbb{C}^{m \times n}$, consider the following four matrix equations for $X \in \mathbb{C}^{n \times m}$:

$$AXA = A,\ XAX = X,\ (AX)^* = AX \text{ and } (XA)^* = XA.$$

There are many ways of showing that these equations have a solution and that such a solution must be unique. Let us assume for the moment that a solution exists, denote this unique solution by $A^\dagger$ and call this *the Moore–Penrose (generalized) inverse of A* [85]. Clearly, if A is nonsingular, then A^{-1} satisfies these four equations and so one has $A^\dagger = A^{-1}$. It is for this reason that the Moore–Penrose inverse is frequently referred to as the *pseudo inverse* or a *generalized inverse*. Note that if $A = 0$, then $A^\dagger = 0^T$. There are several ways of computing the Moore–Penrose inverse. One of the more stable methods is by using the singular value decomposition of the matrix under consideration. Let us suppose that the singular value decomposition

P.N. Shivakumar et al., *Infinite Matrices and Their Recent Applications*,
DOI 10.1007/978-3-319-30180-8_4

of $A \in \mathbb{C}^{m\times n}$ is given by $A = U\Sigma V^*$, where the columns of $U \in \mathbb{C}^{m\times m}$ are mutually orthonormal eigenvectors of AA^*, the columns of $V \in \mathbb{C}^{n\times n}$ are mutually orthonormal eigenvectors of A^*A and $\Sigma \in \mathbb{C}^{m\times n}$ is a block matrix given by

$$\Sigma = \begin{pmatrix} D & 0 \\ 0 & 0 \end{pmatrix}.$$

In the above representation, the matrix $D \in \mathbb{R}^{r\times r}$ is a diagonal matrix whose diagonal entries are given by the positive square roots of the positive eigenvalues of (the hermitian and positive semidefinite matrix) AA^* (or A^*A) and $r = rank(A)$. These entries are referred to as the *singular values of* A. We assume that these singular values are arranged in the decreasing order as they appear in D. It then follows that the Moore–Penrose inverse $\Sigma^\dagger$ of Σ is given by

$$\Sigma^\dagger = \begin{pmatrix} D^{-1} & 0 \\ 0 & 0 \end{pmatrix},$$

where the zero blocks are understood to be the transposes of the appropriate zero blocks of Σ. The expression for $A^\dagger$ is now immediate, viz.,

$$A^\dagger = V\Sigma^\dagger U^*.$$

It may easily be verified that this expression for $A^\dagger$ satisfies the four equations above.

Next, let us look at some formulae. If $0 \neq x \in \mathbb{C}^n$, then $x^\dagger = \frac{1}{\|x\|^2}x^*$, where, of course, the norm is the Euclidean norm. For $A \in \mathbb{C}^{m\times n}$ with full column rank (all the columns are linearly independent), one has

$$A^\dagger = (A^*A)^{-1}A^*,$$

while a dual argument shows that if A has full row rank (all the rows are linearly independent), then

$$A^\dagger = A^*(AA^*)^{-1}.$$

More generally, we have

$$A^\dagger = (A^*A)^\dagger A^* = A^*(AA^*)^\dagger.$$

The following properties are also useful in many computations involving the Moore–Penrose inverse: Let P_M denote the orthogonal projection of $\mathbb{C}^m$ onto a subspace M along $M^\perp$. Then $AA^\dagger = P_{R(A)}$ and $A^\dagger A = P_{R(A^*)}$.

Next, we recall another factorization, which exists for any matrix and is useful in providing an explicit formula for the Moore–Penrose inverse, among other things. For instance, this factorization is also relevant for another type of generalized inverse called the group inverse, which will be discussed a little later.

Let $A \in \mathbb{C}^{m\times n}$ with $r = rank(A) > 0$. Then there exist $F \in \mathbb{C}^{m\times r}$ and $G \in \mathbb{C}^{r\times n}$ such that $A = FG$ and $rank(F) = r = rank(G)$. A factorization defined as above is called a *full-rank factorization* due to the reason that the two factors have full ranks. The full-rank factorization of a matrix is not unique. One way of showing this is by describing a method of obtaining it. Consider a basis (consisting of r vectors in $\mathbb{C}^m$) for $R(A)$. Let F be the matrix of order $m \times r$ whose columns are these r basis vectors. Then $R(F) = R(A)$, by definition and hence $rank(F) = r$. Each column of A is a unique linear combination of the columns of F. The coefficients in each such linear combination are r in number and form a vector in $\mathbb{R}^r$. There are n such vectors. Let G be the matrix whose columns are these n vectors. Then by construction, $A = FG$. It now follows that $rank(G) = r$.

Now, let $A = FG$ be a full-rank factorization so that F has full column rank and G has full row rank. It then follows from the discussion above, that $F^\dagger = (F^*F)^{-1}F^*$ and $G^\dagger = G^*(GG^*)^{-1}$. It now follows that a "reverse order law" for the Moore–Penrose inverse of A holds, viz.,

$$A^\dagger = G^\dagger F^\dagger = G^*(GG^*)^{-1}(F^*F)^{-1}F^*.$$

Note, however, that if $A = BC$, then it is in general, not true that $A^\dagger = C^\dagger B^\dagger$. There is a nice summary for "solutions" of linear equations of the form $Ax = b$, in terms of the Moore–Penrose inverse. Consider the vector $x^* = A^\dagger b$. If the system has a unique solution, it is given by x^*; if it has infinitely many solutions, then the solution with the minimum norm is given by x^*. If the system is inconsistent, then x^* is the least squares solution of the system; if the system has infinitely many least squares solution, then x^* is the least squares solution of the least norm.

Next, we turn our attention to two other types of generalized inverses, one of which always exists and the other type exists only for a specific subclass of matrices. In order to describe these, we need the notion of the index of a square matrix. Let $A \in \mathbb{C}^{n\times n}$. Then the *index of A* is the smallest nonnegative integer k such that $rank(A^k) = rank(A^{k+1})$, if it exists. For a nonsingular matrix A, we set the index to be zero. By considering the range spaces of positive integral powers of A and by using a finite dimensionality argument, one can show that the index exists for any (square) matrix. Now, let k be the index of a singular matrix A such that $r = rank(A^k)$. Then there exist $Q \in \mathbb{C}^{n\times n}$ and $C \in \mathbb{C}^{r\times r}$ both being invertible, such that

$$Q^{-1}AQ = \begin{pmatrix} C & 0 \\ 0 & N \end{pmatrix},$$

where N is a nilpotent matrix of (nilpotent) index k. This is called the *core-nilpotent decomposition* of A. A proof could be found, for instance, in [73]. Given such a decomposition, let us define

$$A^D = Q\begin{pmatrix} C^{-1} & 0 \\ 0 & 0 \end{pmatrix}Q^{-1}.$$

Then the matrix A^D is called the *Drazin inverse* of A [24]. A^D satisfies

$$A^DAA^D = A^D,\ A^DA = AA^D \text{ and } A^{k+1}A^D = A^k.$$

It can be shown that the three equations above with k being the index of A have a unique solution. It now follows that A^D is unique.

Let us now consider the particular case, namely the index being one. So, suppose that A has index 1. Thus $rank(A) = rank(A^2)$. In this case, A^D is denoted by $A^\#$, and is called the *group inverse of* A. We might emphasize that the Drazin inverse exists for all square matrices while the group inverse exists only for square matrices with index 1. In fact, one has the following result (Exercise 5.10.12, [73]): For $A \in \mathbb{C}^{n\times n}$ the following statements are equivalent:

(a) A belongs to a matrix group G.
(b) $R(A) \cap N(A) = 0$.
(c) $R(A)$ and $N(A)$ are complementary subspaces.
(d) index $(A) = 1$.
(e) $Q^{-1}AQ = \begin{pmatrix} C & 0 \\ 0 & 0 \end{pmatrix}$,

for some invertible $Q \in \mathbb{C}^{n\times n}$ and $C \in \mathbb{C}^{r\times r}$, $r = rank(A)$. One has the following formula:

$$A^\# = Q\begin{pmatrix} C^{-1} & 0 \\ 0 & 0 \end{pmatrix}Q^{-1}.$$

An explicit formula for the group inverse also could be given from the full-rank factorization, which in the first place even tells us if the group inverse of A exists. Let us state that result. Let $A = FG$ be a full-rank factorization of A. Then $A^\#$ exists if and only if GF is invertible. In such a case,

$$A^\# = F(GF)^{-2}G,$$

where the superscript -2 denotes "inverse squared."

At this juncture, one could ask if the Moore–Penrose inverse and the group inverse are the same for any class of (square) matrices. In this connection let us recall that a (square) matrix is called *range hermitian*, if $R(A) = R(A^*)$. It is well known [10] that a matrix A is range hermitian if and only if $A^\dagger = A^\#$. As a corollary, if $A \in \mathbb{R}^{n\times n}$ is symmetric, then its Moore–Penrose inverse and the group inverse coincide.

In the final part of this section, we review the notion of the Moore–Penrose inverses of operators in the setting of Hilbert spaces. Our treatment follows the approach that was adopted by Groetsch [41].

Let X, Y be Hilbert spaces and $A \in \mathcal{B}(X, Y)$, where $\mathcal{B}(X, Y)$ is the space of all bounded linear operators from X into Y. We shall restrict our attention to the case when $R(A)$ is a closed subspace of Y. For a given $b \in Y$, consider the operator equation $Ax = b$. If $b \notin R(A)$, then the equation does not have a solution and if $N(A) \neq \{0\}$, then a solution will not be unique, if it exists. Let $P_{R(A)}$ denote the orthogonal projection of Y on to $R(A)$ along $N(A^*)$. Thus, even if the equation above does not have a solution, it seems reasonable to accept any solution of the equation $Ax = P_{R(A)}b$ (which is always consistent) as a generalized solution of $Ax = b$, in some sense. This notion is justified by another approach, this time geometric. Let $u \in X$ be any minimizer of the functional

$$\phi_b(x) = \| Ax - b \|, \ x \in X$$

(assuming that such a minimizer exists). Then it is plausible to call such a u as a generalized solution. In fact these two definitions imply each other and also are equivalent to another condition. We recall this result next.

Theorem 4.1.1 (Theorem 2.1.1 [41]). *Let X, Y be Hilbert spaces and $A \in \mathcal{B}(X, Y)$ with $R(A)$ closed. For $b \in Y$, the following conditions are equivalent:*

(a) $Au = P_{R(A)}b$.
(b) $\| Au - b \| \leq \| Ax - b \|$ *for all* $x \in X$.
(c) $A^*Au = A^*b$.

A vector $u \in X$ which satisfies any of the conditions above is called a *least squares solution* of the equation $Ax = b$. We may now denote the set of all least squares solutions of $Ax = b$ by

$$C_b =: \{u \in X : A^*Au = A^*b\}.$$

Then C_b is a closed and convex set and so it contains a unique vector of minimal norm which we denote by u_b. Define

$$A^\dagger : Y \to X \text{ by } A^\dagger(b) := u_b, \ b \in Y.$$

One can show that $A^\dagger$ is a bounded linear map and that it is the unique solution of the equations: $AXA = A$, $XAX = X$, $(AX)^* = AX$ and $(XA)^* = XA$. As in the matrix case, $A^\dagger$ is called the *Moore–Penrose inverse* of A. We refer the reader to [41] for proofs of these statements and other considerations.

4.2 On the Non-uniqueness of the Moore–Penrose Inverse and Group Inverse of Infinite Matrices

Our intention here is to review some interesting results discovered very recently. Sivakumar [123] gives an example of an invertible infinite matrix V which has infinitely many usual inverses (which are automatically Moore–Penrose inverses) and also has a Moore–Penrose inverse which is not a classical inverse. Thus it follows that there are infinitely many Moore–Penrose inverses of V. In a subsequent work [124], Sivakumar shows that the very same infinite matrix considered earlier has group inverses that are Moore–Penrose inverses, too. These results are interesting in that, for infinite matrices, it follows that the set of generalized inverses is properly larger than the set of usual inverses. It might be remembered that these are not true for finite matrices with real or complex entries. It also follows that for a "symmetric" infinite matrix, the Moore–Penrose inverse need not be the group inverse, unlike the finite symmetric matrix case, where they coincide.

First we give the definitions. It is well known that for infinite matrices multiplication is non-associative. Hence we also impose associativity in the first two Penrose equations (considered in the previous section) and rewrite them as

$$(AX)A = A(XA) = A$$

$$(XA)X = X(AX) = X$$

We call X as a Moore–Penrose inverse of an infinite matrix A if X satisfies the two equations as above and the last two Penrose equations. Similarly, for group inverses of infinite matrices, we demand associativity in the first two equations in its definition.

Suppose that the infinite matrix A can be viewed as a bounded operator between Hilbert spaces. Then it is well known that A has a bounded Moore–Penrose inverse $A^\dagger$ if and only if its range, $R(A)$ is closed [78]. For operators between Banach spaces we have a similar result. However, in such a case, the last two equations involving adjoint should be rewritten in terms of projections on certain subspaces. (See [77] for details.)

In this section, our approach will be purely algebraic, in the sense that infinite matrices are not viewed as operators on some vector spaces. We use the "basis" $\{e^n : n \in \mathbb{N}\}$, where $e^n = (0, 0, \ldots, 1, 0, \ldots)$ with 1 appearing in the nth coordinate. With this notation we consider the infinite matrix V such that $V(e^1) = e^2$ and $V(e^n) = e^{n-1} + e^{n+1}, n \geq 2$. The next result states that V has infinitely many algebraic inverses.

Theorem 4.2.1 (Theorem 0.1 [123]). *Let U and W be infinite matrices defined by*

$$U(e^1) = e^2 - e^4 + e^6 - e^8 + \ldots;\ U(e^2) = e^1$$

and

$$U(e^{n+1}) = e^n - U(e^{n-1}), n \geq 2$$

and

$$W(e^1) = (e^2 - e^4 + e^6 - e^8 + \ldots) - (e^1 - e^3 + e^5 - e^7 + \ldots);\ W(e^2) = e^1$$

and

$$W(e^{n+1}) = e^n - W(e^{n-1}),\ n \geq 2.$$

Then $UV = VU = I, WV = VW = I$ and V has infinitely many algebraic inverses.

The fact that V has a Moore–Penrose inverse which is not a usual inverse is the next result.

Theorem 4.2.2 (Theorem 0.3 [123]). *Let Z be the infinite matrix defined by*

$$Z(e^1) = e^4 - 2e^6 + 3e^8 - 4e^{10} + \cdots,\ Z(e^2) = e^1$$

and

$$Z(e^{n+1}) = e^n - Z(e^{n-1}),\ n \geq 2.$$

Then Z is not a usual inverse of V but Z is a Moore–Penrose inverse of V.

Remark 4.2.3. It follows that the matrix Z defined above satisfies $ZV = I$, whereas, we have $VZ(e^1) \neq e^1$, so that $ZV \neq VZ$. Thus Z is not a group inverse of V. As mentioned earlier, if $A \in \mathbb{R}^{n\times n}$ is symmetric, then its Moore–Penrose inverse and the group inverse coincide. Theorem 4.2.2 shows that the situation is different in the case of infinite matrices. We have a Moore–Penrose inverse that is not a group inverse, even though the infinite matrix V is "symmetric."

The next two results show that there are group inverses of V that also turn out to be Moore–Penrose inverses.

Theorem 4.2.4 (Theorem 3.6 [124]). *For $\alpha \in \mathbb{R}$, let Y_α be the infinite matrix defined by*

$$Y_\alpha(e^1) = \alpha(e^2 - 2e^4 + 3e^6 \cdots) + (e^4 - 2e^6 + 3e^8 - \cdots),$$
$$Y_\alpha(e^2) = \alpha(e^1 - e^3 + e^5 - \cdots) + (e^3 - e^5 + e^7 - \cdots),$$
$$Y_\alpha(e^{2n}) = (-1)^n\{(e^3 - e^5 + \cdots + (-1)^n e^{2n-1}) + (n-1)e^1 - nY_\alpha(e^2)\},\ \textit{for}\ n \geq 2$$

and

$$Y_\alpha(e^{2n+1}) = e^{2n} - Y_\alpha(e^{2n-1}), n \geq 1.$$

For the sake of convenience, let Y denote Y_α. Then Y is a group inverse of V, but not a usual inverse.

Theorem 4.2.5 (Theorem 3.7 [124]). *Let Y be defined as above. Then Y is a Moore–Penrose inverse of V.*

Next, let us briefly review some recent results that were obtained in connection with certain embedding problems. We do not include the precise statements here. Let $M_\infty(\mathbb{Z})$ denote the set of countably infinite square matrices whose entries are integers. Then $M_\infty(\mathbb{Z})$ forms a non-associative groupoid. It is known that every finite or countable groupoid embeds in $M_\infty(\mathbb{Z})$. Roberts and Drazin consider the associated question of $*$-embedding of groupoids with an involution. The main results are proved in (Theorems 2.3 and 3.7, [96]). In the present context, it must be mentioned that these embedding results are used to construct much simpler and more general ways of finding examples like the ones that were constructed by Sivakumar, discussed as above. We refer the reader to the details in [96].

In this last part, we review the work of Campbell [18] on the Drazin inverses of infinite matrices. First, we recall some definitions. Let A be an infinite matrix. An infinite matrix X is referred to as a C-(2) inverse of A if

$$AX = XA \text{ and } X(AX) = (XA)X = X.$$

X is called a C-(1,2) inverse of A if X is a C-(2) inverse of A and

$$A(XA) = (AX)A = A.$$

Let X, Y be C-(2) inverses of A such that

$$X(AY) = (XA)Y = X$$

and

$$Y(AX) = (YA)X = X.$$

Then we write $X \subseteq Y$. The two conditions given above may be described in words, as follows. The "null space" of Y is contained in the "null space" of X and the "range space" of X is contained in the "range space" of Y. If $X \subseteq Y$ for all C-(2) inverses X, we call Y maximal. The author proposes the following definition: A unique maximal C-(2) inverse for an infinite matrix A, if it exists, is called the Drazin inverse of A. It is denoted by A^D. If A^D exists and is a C-(1,2) inverse of A, then it is also called the group inverse of A and is denoted by $A^{\#}$. For finite matrices, these reduce to the corresponding definitions given earlier. It is shown that (Corollary 2, [18]) if A has a maximal C-(2) inverse then A^D exists. The author also discusses applications to Markov chains and infinite systems of differential equations.

Chapter 5
Generalized Inverses: Quaternions

5.1 Introduction

A quaternion algebra $\mathbb{H}$ was discovered by Sir Rowan Hamilton in 1843, which is a four-dimensional non-commutative algebra over real number field $\mathbb{R}$ with canonical basis $\{1, \mathbf{i}, \mathbf{j}, \mathbf{k}\}$ satisfying the conditions:

$$\mathbf{i}^2 = \mathbf{j}^2 = \mathbf{k}^2 = \mathbf{ijk} = -1,$$

so that one has

$$\mathbf{ij} = -\mathbf{ji} = \mathbf{k}, \ \mathbf{jk} = -\mathbf{kj} = \mathbf{i}, \ \text{ and } \ \mathbf{ki} = -\mathbf{ik} = \mathbf{j}.$$

Any element $\alpha \in \mathbb{H}$ can be written in a unique way: $\alpha = a + b\mathbf{i} + c\mathbf{j} + d\mathbf{k}$, where a, b, c, and d are real numbers, i.e., $\mathbb{H} = \{a + b\mathbf{i} + c\mathbf{j} + d\mathbf{k} \mid a, b, c, d \in \mathbb{R}\}$. The conjugate of α is defined as $\bar{\alpha} = a - b\mathbf{i} - c\mathbf{j} - d\mathbf{k}$, and the norm $|\alpha|$ is given by $|\alpha| = \sqrt{\alpha\bar{\alpha}}$. It is well-known that $\mathbb{H}$ is a skew field (or called a division ring).

The study of polynomials with quaternion coefficients may go back to Niven [79, 80] in the early 1940s. In these two seminal papers, Niven established the "Fundamental Theorem of Algebras" for quaternions, that is, $x^m + a_1x^{m-1} + a_2x^{m-2} + \cdots + a_m = 0$ $(a_m \neq 0)$ with coefficients in a division ring D has a solution in D if and only if the centre C of D is a real-closed field and D is the algebra of real quaternions over C. Furthermore, Niven proved that there may be infinitely many roots or a finite number, but in the latter case there are at most $(2m-1)^2$, which shows the essential difference between the polynomials over division rings and over commutative fields.

Unlike the polynomials over commutative fields, there are several forms of quaternion polynomials depending on the positions of coefficients due to the non-commutativity of $\mathbb{H}$. For example, regular quaternion polynomials in [22] and

P.N. Shivakumar et al., *Infinite Matrices and Their Recent Applications*,
DOI 10.1007/978-3-319-30180-8_5

quaternion simple polynomials in [81]. Some properties of these polynomials have been discussed (see, for example, [63, 86]). In this chapter, we will use the following Definition 5.1.1, which places the coefficients on the left side of a variable x:

Definition 5.1.1. A quaternion polynomial $f(x)$ over $\mathbb{H}$ is defined as

$$f(x) = a_n x^n + \cdots + a_1 x + a_0, \quad a_i \in \mathbb{H}, a_n \neq 0, \ i = 0, \ldots, n,$$

where x commutes with each element in $\mathbb{H}$.

The set of all quaternion polynomials in x is denoted by $\mathbb{H}[x]$. The degrees, leading terms, and leading coefficients are defined in a natural way. It is well-known that $\mathbb{H}[x]$ becomes a non-commutative domain under the usual polynomial operations.

The quaternion polynomials and matrices with quaternion polynomial entries have been widely studied with many applications in the past decades. For example, in [145], the Fast Fourier Transform for the product of two quaternion polynomial has been discussed with the complexity analysis. In [22], they studied Gröbner basis theory for the ring of quaternion polynomials and explored how to compute the module syzygy. Smith-McMillan forms of quaternion polynomial matrices are defined and some applications to dynamical systems are given in [87]. Some properties of Ore matrices can be found in [37, 144].

For matrices over commutative rings, it is well-known that the various generalized inverses have been defined and explored for many years (see, for example, [10, 85]). This motivates us to consider the generalized inverses questions for quaternion polynomial matrices. The numerical computations for generalized inverses have been discussed for a long time. We will use the symbolic computational methods which have attracted more and more attentions recently, for example, [42, 45, 46, 89].

The structure of this chapter is as follows. In Sect. 5.2, we discuss $\{1\}$-inverses of the quaternion polynomial matrices and present an algorithm to determine the existence of $\{1\}$-inverses. Using one-sided greatest common divisors of quaternion polynomials, we develop an efficient algorithm to compute $\{1\}$-inverses if they exist. In Sect. 5.3, we give the definition of the Moore–Penrose inverse for quaternion polynomial matrices and discuss some basic properties. This includes a necessary and sufficient condition for the existence of the Moore–Penrose inverse. In Sect. 5.4, the well-knownv Leverrier–Faddeev algorithm is extended to quaternion polynomial matrices by using generalized characteristic polynomials. Finally, we discuss the interpolation problems for quaternion polynomials and give an efficient algorithm to compute the Moore–Penrose inverse in Sect. 5.5. We have implemented our algorithms in Maple and some examples are also given in Sect. 5.6.

5.2 {1}-Inverses of Quaternion Matrices

In this section, we first discuss some properties of {1}-inverses of quaternion polynomial matrices, which will be used to formulate an algorithm for finding {1}-inverses for quaternion polynomial matrices. Let $\mathbb{H}[x]^{m\times n}$ be the set of all $m \times n$ matrices over $\mathbb{H}[x]$. Recall that $A \in \mathbb{H}[x]^{m\times n}$ has a {1}-inverse $G \in \mathbb{H}[x]^{n\times m}$ if $AGA = A$.

The main technical idea here is to use a well-known result: $\mathbb{H}[x]$ is a non-commutative principal ideal domain. From this point, we can define one-sided greatest common divisors and least common multiples of quaternion polynomials as follows:

Let $f, g \in \mathbb{H}[x]\backslash\{0\}$. A *greatest common right divisor (GCRD)* of f and g, written $\mathrm{gcrd}(f, g)$, is a nonzero monic $d \in \mathbb{H}[x]$ such that

(a) d is a common right divisor of f and g, namely $f = f_1 d, g = g_1 d$ for some $f_1, g_1 \in \mathbb{H}[x]$;
(b) If $d_1 \in \mathbb{H}[x]$ is a common right divisor of f and g, then d_1 is a right divisor of d.

A *least common right multiple (LCRM)* of f and g, written $\mathrm{lcrm}(f, g)$, is a nonzero monic $s \in \mathbb{H}[x]$ such that

(1) s is a common right multiple of f and g, namely $s = ff_1 = gg_1$ for some $f_1, g_1 \in \mathbb{H}[x]$,
(2) If $s_1 \in \mathbb{H}[x]$ is a common right multiple of f and g, then s_1 is a right multiple of s.

It is easy to prove that GCRD and LCRM are unique. The *greatest common left divisor (GCLD)* and the *least common left multiple* (LCLM) of f and g are defined correspondingly. The following two lemmas can be proved by using the properties of one-sided principal ideals.

Lemma 5.2.1. *Let $a_1, a_2, \ldots, a_n, d \in \mathbb{H}[x]$ and d be monic. The following statements are equivalent:*

(i) $\mathbb{H}[x]a_1 + \mathbb{H}[x]a_2 + \cdots + \mathbb{H}[x]a_n = \mathbb{H}[x]d$.
(ii) $d = \mathrm{gcrd}(a_1, a_2, \ldots, a_n)$.

Lemma 5.2.2. *Let $a_1, a_2, \ldots, a_n, s \in \mathbb{H}[x]$ and s be monic. The following statements are equivalent:*

(i) $a_1\mathbb{H}[x] \bigcap a_2\mathbb{H}[x] \bigcap \cdots \bigcap a_n\mathbb{H}[x] = s\mathbb{H}[x]$.
(ii) $s = \mathrm{lclm}(a_1, a_2, \ldots, a_n)$.

There are several ways to compute the GCRD and LCLM (see, for example, [22]). Here we use the following algorithm that is analogous to the traditional extended Euclidean algorithm for commutative Euclidean domain ([135], Algorithm 3.6). For $f = qg + r$, we denote $q := f \,\mathrm{quo}_l\, g$ the left quotient of the division of f by g.

Algorithm 1 Extended Euclidean Algorithm (EEA)

Input $f, g \in \mathbb{H}[x]$, where $\deg(f) = n$, $\deg(g) = m$, $m \leq n$, $m, n \in \mathbb{N}$.
Output $k \in \mathbb{N}$, $r_i, s_i, t_i \in \mathbb{H}[x]$ for $0 \leq i \leq k+1$, and $q_i \in \mathbb{H}[x]$ for $1 \leq i \leq k$, as computed below.

1: $r_0 \leftarrow f$, $s_0 \leftarrow 1$, $t_0 \leftarrow 0$, $r_1 \leftarrow g$, $s_1 \leftarrow 0$, $t_1 \leftarrow 1$
2: $i \leftarrow 1$
3: **while** $r_i \neq 0$ **do**
    $q_i \leftarrow r_{i-1} \operatorname{quo}_l r_i$, $r_{i+1} \leftarrow r_{i-1} - q_i r_i$,
    $s_{i+1} \leftarrow s_{i-1} - q_i s_i$, $t_{i+1} \leftarrow t_{i-1} - q_i t_i$, $i \leftarrow i+1$.
4: **end while**
5: $k \leftarrow i - 1$
6: **return** k, r_i, s_i, t_i for $0 \leq i \leq k+1$, and q_i for $1 \leq i \leq k$.

The correctness of the above algorithm follows the strictly decreasing degrees: $\deg(r_1) > \deg(r_2) > \cdots > \deg(r_k) \geq 0$. Next, we shall verify that above algorithm also produces some one-sided greatest common divisors and least common multiples in quaternion polynomial case, which we shall use later.

Lemma 5.2.3. *Let r_i, s_i, t_i for $0 \leq i \leq k+1$ and q_i for $1 \leq i \leq k$ be as in Algorithm 1. Consider the matrices*

$$R_0 = \begin{bmatrix} s_0 & t_0 \\ s_1 & t_1 \end{bmatrix}, \; Q_i = \begin{bmatrix} 0 & 1 \\ 1 & -q_i \end{bmatrix} \; \textit{for } 1 \leq i \leq k$$

in $\mathbb{M}_{2\times 2}(\mathbb{H}[x])$, and $R_i = Q_i \cdots Q_1 R_0$ for $0 \leq i \leq k$. Then

(a) $R_i \begin{bmatrix} f \\ g \end{bmatrix} = \begin{bmatrix} r_i \\ r_{i+1} \end{bmatrix}$.
(b) $R_i = \begin{bmatrix} s_i & t_i \\ s_{i+1} & t_{i+1} \end{bmatrix}$.
(c) $s_i f + t_i g = r_i$ *for all* $1 \leq i \leq k+1$.
(d) $\operatorname{gcrd}(f, g) = r_k$.
(e) $\operatorname{lclm}(f, g) = s_{k+1} f = t_{k+1} g$.

Proof. (a) and (b) can be proved by mathematical induction on i and the relation $R_i = Q_i R_{i-1}$. (c) follows directly from (a).

To prove (d), from assumptions and (a)–(c), we have

$$\begin{bmatrix} r_k \\ 0 \end{bmatrix} = R_k \begin{bmatrix} f \\ g \end{bmatrix} = Q_k \cdots Q_1 R_0 \begin{bmatrix} f \\ g \end{bmatrix} = Q_k \cdots Q_1 \begin{bmatrix} r_0 \\ r_1 \end{bmatrix} = Q_k \cdots Q_1 \begin{bmatrix} f \\ g \end{bmatrix}.$$

Note that for each $i \in \{1, \ldots, k\}$, Q_i has an invertible $Q_i^{-1} = \begin{bmatrix} q_i & 1 \\ 1 & 0 \end{bmatrix}$ over $\mathbb{H}[x]$.

Hence

$$\begin{bmatrix} f \\ g \end{bmatrix} = Q_1^{-1} \cdots Q_k^{-1} \begin{bmatrix} r_k \\ 0 \end{bmatrix},$$

which implies that r_k is a common right divisor of f and g. On the other hand, by (c), $r_k = s_k f + t_k g$ implies that any common right divisor of f and g is also a right divisor of r_k. Therefore, $\mathrm{gcrd}(f, g) = r_k$.

Finally, since $0 = r_{k+1} = s_{k+1} f + t_{k+1} g$, we have $h = s_{k+1} f = -t_{k+1} g$ is a left common multiple of f and g. Meanwhile, $\deg(h) = \deg(f) + \deg(g) - \deg(\mathrm{gcrd}(f, g))$. Thus, $h = \mathrm{lclm}(f, g)$. □

Our purpose is to use one-sided greatest common divisors and least common multiples to compute {1}-inverses of quaternion polynomial matrices. Our algorithm is based on recursively computing GCRDs and LCLMs. The following result is well-known in commutative case.

Theorem 5.2.4. *Let* $A = \begin{bmatrix} a & \vec{b} \\ \vec{0} & B \end{bmatrix} \in \mathbb{H}[x]^{(m+1)\times(n+1)}$ *with* $0 \neq a \in \mathbb{H}[x]$, $\vec{b} = [b_1 \cdots b_n] \in \mathbb{H}[x]^{1\times n}$ *and* $B \in \mathbb{H}[x]^{m\times n}$.

(a) If A has a {1}*-inverse over* $\mathbb{H}[x]$, *then* $\mathrm{gcld}(a, b_1, \ldots, b_n) = 1$.

(b) Suppose $A = \begin{bmatrix} a & \vec{0} \\ \vec{0} & B \end{bmatrix}$. *If A has a* {1}*-inverse over* $\mathbb{H}[x]$, *then* $a \in \mathbb{H}$ *and B has a* {1}*-inverse over* $\mathbb{H}[x]$.

Proof. Let $G = \begin{bmatrix} g & \vec{h} \\ \vec{k} & H \end{bmatrix}$ be a {1}-inverse of A, where $g \in \mathbb{H}$, $\vec{h} = [h_1, \ldots, h_n] \in \mathbb{H}[x]^{1\times n}$, $\vec{k} = [k_1 \cdots k_m]^T \in \mathbb{H}[x]^{m\times 1}$ and $H \in \mathbb{H}[x]^{n\times m}$.

Since $A = AGA$, we have

$$\begin{bmatrix} a & \vec{b} \\ \vec{0} & B \end{bmatrix} = \begin{bmatrix} a & \vec{b} \\ \vec{0} & B \end{bmatrix} \begin{bmatrix} g & \vec{h} \\ \vec{k} & H \end{bmatrix} \begin{bmatrix} a & \vec{b} \\ \vec{0} & B \end{bmatrix} = \begin{bmatrix} aga + \vec{b}\vec{k}a & * \\ * & B\vec{k}\vec{b} + BHB \end{bmatrix}. \tag{5.1}$$

Then $aga + \vec{b}\vec{k}a = a$, and thus $(ag + \vec{b}\vec{k} - 1)a = 0$. Since $\mathbb{H}[x]$ is a principal ideal domain, we have $ag + \vec{b}\vec{k} - 1 = 0$, i.e., $ag + b_1k_1 + \cdots + b_nk_n = 1$. Therefore, $\mathrm{gcld}(a, b_1, \ldots, b_n) = 1$.

To prove (b), let $\vec{b} = 0$ in (5.1). Comparing the correspondent entries of matrices on both sides, we have $aga = a$ and $BHB = B$. Hence B has a {1}-inverse H over $\mathbb{H}[x]$. $aga = a$ implies that $(ag - 1)a = 0$, and either $a = 0$ or $ag = 1$ since $\mathbb{H}[x]$ is a domain. Note that both a and g are quaternion polynomials. Therefore $a \in \mathbb{H}$. □

Corollary 5.2.5. *Let* $A = \begin{bmatrix} 1 & \vec{0} \\ \vec{0} & B \end{bmatrix} \in \mathbb{H}[x]^{(m+1)\times(n+1)}$. *Then A has a* $\{1\}$*-inverse over* $\mathbb{H}[x]$ *if and only if B a* $\{1\}$*-inverse over* $\mathbb{H}[x]$*. Moreover, if* $C \in \mathbb{H}[x]^{n\times m}$ *is a* $\{1\}$*-inverse of B, then* $\begin{bmatrix} 1 & \vec{0} \\ \vec{0} & C \end{bmatrix}$ *is a* $\{1\}$*-inverse of A over* $\mathbb{H}[x]$*.*

Proof. If A has a $\{1\}$-inverse over $\mathbb{H}[x]$, then by Theorem 5.2.4(b), B has a $\{1\}$-inverse over $\mathbb{H}[x]$. Conversely, suppose $C \in \mathbb{H}[x]^{n\times m}$ is a $\{1\}$-inverse of B, that is, $BCB = B$. Let $G = \begin{bmatrix} 1 & \vec{0} \\ \vec{0} & C \end{bmatrix}$. Then

$$AGA = \begin{bmatrix} 1 & \vec{0} \\ \vec{0} & BCB \end{bmatrix} = \begin{bmatrix} 1 & \vec{0} \\ \vec{0} & B \end{bmatrix} = A,$$

and so G is a $\{1\}$-inverse of A over $\mathbb{H}[x]$. Therefore, A has a $\{1\}$-inverse over $\mathbb{H}[x]$. □

Next we shall discuss the row and column transformations of quaternion polynomial matrices. Unlike matrices over fields, we cannot use the usual three elementary row (column) transformations freely since $\mathbb{H}[x]$ is a non-commutative domain, not a field. In the following, we will show how to use one-sided greatest common divisors and least common multiples to make row (column) transformations.

Lemma 5.2.6. *Let* $E, E_1 \in \mathbb{H}[x]^{n\times n}$. *Then* $EE_1 = I$ *implies* $E_1E = I$.

Proof. Since $\mathbb{H}[x]$ is a principal ideal domain and Noetherian, we know that $\mathbb{H}[x]$ is stably finite by Proposition 1.13 in [62]. Hence E_1 is also a left inverse of E. □

Lemma 5.2.7. *Let* $A = \begin{bmatrix} a_{11} & a_{12} \\ a_{21} & a_{22} \end{bmatrix} \in \mathbb{H}[x]^{2\times 2}$, $g_R = \mathrm{gcrd}(a_{11}, a_{21})$ *and* $g_L = \mathrm{gcld}(a_{11}, a_{12})$. *Then there exist invertible matrices* $E, F \in \mathbb{H}[x]^{2\times 2}$, *such that*

$$EA = \begin{bmatrix} g_R & * \\ 0 & * \end{bmatrix}, \qquad AF = \begin{bmatrix} g_L & 0 \\ * & * \end{bmatrix},$$

where each $*$ *stands for some element in* $\mathbb{H}[x]$.

Proof. To design our algorithms, the following proof is constructive. If only one of a_{11} and a_{21} is equal to zero, then we just simply switch two rows. If both of a_{11} and a_{21} are equal to zero, then we will do nothing. Now we assume that both of a_{11} and a_{21} are nonzero. Using the Extended Euclidean Algorithm 1 and Lemmas 5.2.3, we can calculate $s, t, k, l \in \mathbb{H}[x]$, such that

$$sa_{11} + ta_{21} = g_R, \qquad \mathrm{lclm}(a_{11}, a_{21}) = ka_{11} = la_{21}. \tag{5.2}$$

Compute $b_{11}, b_{21} \in \mathbb{H}[x]$ such that $a_{11} = b_{11}g_R$, $a_{21} = b_{21}g_R$. Then $(sb_{11} + tb_{21} - 1)g_R = 0$, and $(kb_{11} - lb_{21})g_R = 0$. Since $\mathbb{H}[x]$ is a domain and $g_R \neq 0$, we have

$$sb_{11} + tb_{21} = 1 \qquad \text{and} \qquad kb_{11} - lb_{21} = 0. \tag{5.3}$$

Again, by (5.2), $\text{gcld}(k, l) = 1$, and we can use the Extended Euclidean Algorithm 1 to find $p, q \in \mathbb{H}[x]$ such that

$$kp - lq = 1. \tag{5.4}$$

Now set

$$E = \begin{bmatrix} s & t \\ k & -l \end{bmatrix}, \qquad E_1 = \begin{bmatrix} b_{11} & p - b_{11}sp - b_{11}tq \\ b_{21} & q - b_{21}sp - b_{21}tq \end{bmatrix}.$$

Then, by (5.2)–(5.4),

$$EA = \begin{bmatrix} s & t \\ k & -l \end{bmatrix} \begin{bmatrix} a_{11} & a_{12} \\ a_{21} & a_{22} \end{bmatrix} = \begin{bmatrix} sa_{11} + ta_{11} & * \\ ka_{11} - la_{21} & * \end{bmatrix} = \begin{bmatrix} g_R & * \\ 0 & * \end{bmatrix}$$

and

$$EE_1 = \begin{bmatrix} s & t \\ k & -l \end{bmatrix} \begin{bmatrix} b_{11} & p - b_{11}sp - b_{11}tq \\ b_{21} & q - b_{21}sp - b_{21}tq \end{bmatrix}$$

$$= \begin{bmatrix} sb_{11} + tb_{21} & s(p - b_{11}sp - b_{11}tq) + t(q - b_{21}sp - b_{21}tq) \\ kb_{11} - lb_{21} & k(p - b_{11}sp - b_{11}tq) - l(q - b_{21}sp - b_{21}tq) \end{bmatrix}$$

$$= \begin{bmatrix} 1 & sp - (sb_{11} + tb_{21})sp - (sb_{11} + tb_{21})tq + tq \\ 0 & kp - lq - (kb_{11} - lb_{21})sp - (kb_{11} - lb_{21})tq \end{bmatrix}$$

$$= \begin{bmatrix} 1 & 0 \\ 0 & 1 \end{bmatrix}.$$

Thus, E_1 is a right inverse of E over $\mathbb{H}[x]$. By Lemma 5.2.6, E_1 is also a left inverse of E.

The construction for F can be done in a similar way. □

Next we generalize this kinds of row/column transformations determined by one-sided greatest common divisors and least common multiplies to matrices with arbitrary sizes.

Theorem 5.2.8. *Let $A = (a_{ij}) \in \mathbb{H}[x]^{m\times n}$. Then we can compute two invertible matrices, $E \in \mathbb{H}[x]^{m\times m}$ and $F \in \mathbb{H}[x]^{n\times n}$, such that*

$$EA = \begin{bmatrix} g_R & * \\ \vec{0} & * \end{bmatrix}, \qquad AF = \begin{bmatrix} g_L & \vec{0} \\ * & * \end{bmatrix},$$

where $g_R = \operatorname{gcrd}(a_{11}, \dots, a_{m1})$, $g_L = \operatorname{gcld}(a_{11}, \dots, a_{1n})$, and each $*$ stands for some matrix with suitable size over $\mathbb{H}[x]$.

Proof. Using Lemma 5.2.7, we can compute an invertible matrix $E_1 \in \mathbb{H}[x]^{2\times 2}$ such that

$$\begin{bmatrix} E_1 & \vec{0} \\ \vec{0} & I_{m-2} \end{bmatrix} A = \begin{bmatrix} \operatorname{gcrd}(a_{11}, a_{21}) & * & \cdots & * \\ 0 & * & \cdots & * \\ a_{31} & a_{32} & \cdots & a_{3n} \\ \vdots & \vdots & \ddots & \vdots \\ a_{m1} & a_{m2} & \cdots & a_{mn} \end{bmatrix}.$$

It is easy to see that $\begin{bmatrix} E_1 & \vec{0} \\ \vec{0} & I_{m-2} \end{bmatrix}$ is invertible over $\mathbb{H}[x]$. If $a_{31} = 0$, we will go to a_{41}. Otherwise, interchange row 2 and row 3 by multiplying an elementary matrix M on the left side. Then applying Lemma 5.2.7 to the 2×2-matrix on the upper left corner to compute an invertible matrix $E_2 \in \mathbb{H}[x]^{2\times 2}$ such that

$$\begin{bmatrix} E_2 & \vec{0} \\ \vec{0} & I_{m-2} \end{bmatrix} M \begin{bmatrix} E_1 & \vec{0} \\ \vec{0} & I_{m-2} \end{bmatrix} A = \begin{bmatrix} \operatorname{gcrd}(a_{11}, a_{21}, a_{31}) & * & \cdots & * \\ 0 & * & \cdots & * \\ 0 & * & \cdots & * \\ a_{41} & a_{42} & \cdots & a_{4n} \\ \vdots & \vdots & \ddots & \vdots \\ a_{m1} & a_{m2} & \cdots & a_{mn} \end{bmatrix}.$$

Again, it is easy to verify that $\begin{bmatrix} E_2 & \vec{0} \\ \vec{0} & I_{m-2} \end{bmatrix} M \begin{bmatrix} E_1 & \vec{0} \\ \vec{0} & I_{m-2} \end{bmatrix}$ is invertible over $\mathbb{H}[x]$. Continuing on the same process, we can obtain an invertible matrix $E \in \mathbb{H}[x]^{m\times m}$, such that $EA = \begin{bmatrix} g_R & * \\ \vec{0} & * \end{bmatrix}$.

The construction of the matrix F can be done in a similar way. □

Based on the above results, we design an algorithm for computing a $\{1\}$-inverse of a given matrix over the quaternion polynomial ring $\mathbb{H}[x]$.

Algorithm 2 Computing a {1}-inverse of a given matrix over $\mathbb{H}[x]$

Input $A = (a_{ij}) \in \mathbb{H}[x]^{m\times n}$.

Output $\begin{cases} \text{a } \{1\}\text{-inverse of } G \in \mathbb{H}[x]^{n\times m} \text{ such that } AGA = A, \\ \text{"no } \{1\}\text{-inverse exist.", otherwise} \end{cases}$

1: Computing $g_1 \leftarrow \text{gcrd}(a_{11}, a_{21}, \ldots, a_{m1})$

2: Computing an invertible matrices $E \in \mathbb{H}[x]^{m\times m}$ such that

$$EA = \begin{bmatrix} g_1 & \vec{b} \\ \vec{0} & * \end{bmatrix}, \text{ where } \vec{b} = \begin{bmatrix} b_1 & \cdots & b_{n-1} \end{bmatrix}.$$

3: Computing $g_2 \leftarrow \text{gcld}(g_1, b_1, \ldots, b_{n-1})$ and an invertible matrix $F \in \mathbb{H}[x]^{n\times n}$ such that

$$(EA)F = \begin{bmatrix} g_2 & \vec{0} \\ * & B \end{bmatrix},$$

4: **if** $g_2 \neq 1$ **then return** "no {1}-inverse exist."

5: **else** use usual column transformations and computing an invertible matrix $M \in \mathbb{H}[x]^{m\times m}$ such that

$$M((EA)F) = \begin{bmatrix} 1 & \vec{0} \\ \vec{0} & B \end{bmatrix}$$

Recursively call Algorithm 2 to determine (compute) if B has a {1}-inverse. If find a {1}-inverse H of B over $\mathbb{H}[x]$,

return $G \leftarrow F \begin{bmatrix} 1 & \vec{0} \\ \vec{0} & H \end{bmatrix} ME$

6: **end if**

Theorem 5.2.9. *Algorithm 2 is correct.*

Proof. Note that

$$MEAF = \begin{bmatrix} 1 & \vec{0} \\ \vec{0} & B \end{bmatrix}, \qquad G = F \begin{bmatrix} 1 & \vec{0} \\ \vec{0} & H \end{bmatrix} ME, \qquad BHB = B.$$

We have

$$(MEA)G(AF) = MEAF \begin{bmatrix} 1 & \vec{0} \\ \vec{0} & H \end{bmatrix} MEAF$$

$$= \begin{bmatrix} 1 & \vec{0} \\ \vec{0} & B \end{bmatrix} \begin{bmatrix} 1 & \vec{0} \\ 0_{(n-1)\times 1} & H \end{bmatrix} \begin{bmatrix} 1 & \vec{0} \\ \vec{0} & B \end{bmatrix}$$

$$= \begin{bmatrix} 1 & \vec{0} \\ \vec{0} & BHB \end{bmatrix}$$

$$= \begin{bmatrix} 1 & \vec{0} \\ \vec{0} & B \end{bmatrix}$$

$$= MEAF.$$

Since E, F, and M are invertible over $\mathbb{H}[x]$ (Theorem 5.2.8), we have $AGA = A$, which completes the proof. □

5.3 The Moore–Penrose Inverse

It is well-known that the Moore–Penrose inverse is the most famous generalized inverse with numerous applications. In the following sections, we discuss the Moore–Penrose inverse for matrices over $\mathbb{H}[x]$.

The conjugate of $f(x) = a_n x^n + \cdots + a_0 \in \mathbb{H}[x]$ is defined as $\bar{f}(x) = \bar{a}_n x^n + \cdots + \bar{a}_0$. For $A \in \mathbb{H}[x]^{m\times n}$, the conjugate $\bar{A}$ of A is defined as $\bar{A} = (\overline{A_{ij}})$. Moreover, $A^T, A^* \in \mathbb{H}[x]^{n\times m}$ denote the transpose and the conjugate transpose of A, respectively. More properties can be found in, for example, [87, 88].

Lemma 5.3.1 ([88]). *Let $f, g \in \mathbb{H}[x]$. Then (i) $\overline{fg} = \bar{g}\bar{f}$ (ii) $f\bar{f} = \bar{f}f \in \mathbb{R}[x]$ (iii) If $fg \in \mathbb{R}[x]$, then $fg = gf$.*

Definition 5.3.2. A matrix in $\mathbb{H}[x]^{n\times m}$ is called a Moore–Penrose inverse of $A \in \mathbb{H}[x]^{m\times n}$ if it is a solution of the following system of equations:

$$AXA = A,\ XAX = X,\ (AX)^* = AX,\ (XA)^* = XA.$$

It is easy to prove that if there is a solution, then it is unique. As usual, we denote the Moore–Penrose inverse of A as $A^\dagger$. Using similar methods in commutative case, we can get some properties for quaternion polynomial matrices, for example:

Proposition 5.3.3. *Let $A \in \mathbb{H}[x]^{m\times n}$ with $A^\dagger$. Then*

(i) $(A^*)^\dagger = (A^\dagger)^*$, $A^\dagger (A^\dagger)^* A^* = A^\dagger = A^* (A^\dagger)^* A^\dagger$ *and* $A^\dagger A A^* = A^* = A^* A A^\dagger$.
(ii) Let $U \in \mathbb{H}^{m\times m}$ is a unitary matrix, that is, $UU^* = U^*U = I_m$. Then $(UA)^\dagger = A^\dagger U^*$.

Lemma 5.3.4. *If $E \in \mathbb{H}[x]^{m\times m}$ and satisfies $E = E^2 = E^*$, then $E \in \mathbb{H}^{m\times m}$.*

Proof. Let $f_1, \ldots, f_m$ be the entries on the first row of E. From $E = E^*$, without loss of generality, we may assume that $f_1 = \bar{f_1} \neq 0$. Then by $E = E^2$, we have

$$f_1 = f_1\bar{f_1} + \sum_{i=2}^{m} f_i\bar{f_i} = f_1^2 + \sum_{i=2}^{m} f_i\bar{f_i}.$$

Since $f_1 = \overline{f_1}$, the leading coefficient of f_1^2 is a positive real number. Note that the leading coefficient of $\sum_{i=2}^{m} f_i\overline{f_i}$ is also a positive real number. Thus,

$$\deg\left(f_1^2\right) \geq \deg f_1 = \deg\left(f_1^2 + \sum_{i=2}^{m} f_i\overline{f_i}\right)$$

$$= \max\left\{\deg\left(f_1^2\right), \deg\left(\sum_{i=2}^{m} f_i\overline{f_i}\right)\right\} \geq \deg\left(f_1^2\right).$$

This shows that $f_1 \in \mathbb{H}$. Furthermore, $0 = \deg f_1 = \deg\left(\sum_{i=2}^{m} f_i\overline{f_i}\right)$ and the leading coefficients of $\{f_i\bar{f_i}\}$ $(f_i \neq 0)$ are positive reals imply that $f_i \in \mathbb{H}$ for all $1 \leq i \leq m$. The same discussion can be done for the other rows of E. Therefore, $E \in \mathbb{H}^{m\times m}$. □

Note that we require that $A^\dagger$ must be in $\mathbb{H}[x]^{n\times m}$. Therefore unlike matrices over fields or skew fields, the Moore–Penrose inverses for some quaternion polynomial matrices might not exist. Clearly, $A^\dagger$ must be a $\{1\}$-inverse of A. Thus algorithms in Sect. 5.2 provide a way to check that $A^\dagger$ doesn't exist. In general, we don't have efficient algorithms to verify the existence of $A^\dagger$.

Next we will give conditions for quaternion polynomial matrices to have Moore–Penrose inverses. But the proofs are non-constructive.

It is easy to see that $A \in \mathbb{H}^{m\times n}[x]$ induces an additive homomorphism from $\mathbb{H}[x]^{n\times 1}$ to $\mathbb{H}[x]^{m\times 1}$, that is, for all $P, Q \in \mathbb{H}[x]^{n\times 1}$, $A(P+Q) = AP + AQ \in \mathbb{H}[x]^{m\times 1}$. By the definition of Moore–Penrose inverses and Proposition 5.3.3, it is easy to prove the following lemma:

Lemma 5.3.5. *Let $A \in \mathbb{H}[x]^{m\times n}$ such that $A^\dagger$ exists. Considering A as a homomorphism from $\mathbb{H}[x]^{n\times 1}$ to $\mathbb{H}[x]^{m\times 1}$, one has $Image(A) = Image(AA^*) = Image(AA^\dagger)$ and $Image(A^*) = Image(A^*A) = Image(A^\dagger A)$.*

It is well-known that there are two types of eigenvalues for a given quaternion matrix $A_{m\times n}$: right eigenvalues and left eigenvalues, since $\mathbb{H}$ is a non-commutative domain. Right eigenvalues have been studied extensively (see, for example, [4, 14, 64]). We shall work with right eigenvalues towards our main result, that is, find a nonzero vector $\vec{x} \in \mathbb{H}^{n\times 1}$ and a $\lambda \in \mathbb{H}$ such that $A\vec{x} = \vec{x}\lambda$. For simplicity, we shall just use the term "eigenvalue" instead of right eigenvalue from now on. The following result is well-known and very useful.

Lemma 5.3.6 ([141]). *$A \in \mathbb{H}^{m\times m}$ is hermitian, that is, $A = A^*$, if and only if there exists a unitary matrix $U \in \mathbb{H}^{m\times m}$ such that $U^*AU = diag(\lambda_1, \ldots, \lambda_m)$, where λ_i are the eigenvalues of A.*

Now we are ready to give conditions that quaternion polynomial matrices must satisfy in order to have Moore–Penrose inverses. The following theorem is well-known in some cases, see, for example, [10, 91]. Here is an analogue for quaternion polynomial matrices.

Theorem 5.3.7. *Let* $A \in \mathbb{H}[x]^{m\times n}$. *Then* $A^\dagger$ *exists if and only if* $A = U\begin{pmatrix} A_1 & A_2 \\ 0 & 0 \end{pmatrix}$ *with* $U \in \mathbb{H}^{m\times m}$ *unitary and* $A_1A_1^* + A_2A_2^*$ *a unit in* $\mathbb{H}[x]^{r\times r}$ *with* $r \le \min\{m, n\}$. *Moreover,*

$$A^\dagger = \begin{pmatrix} A_1^*\left(A_1A_1^* + A_2A_2^*\right)^{-1} & 0 \\ A_2^*\left(A_1A_1^* + A_2A_2^*\right)^{-1} & 0 \end{pmatrix} U^*.$$

Proof. ($\Longrightarrow$) If A has the Moore–Penrose inverse $A^\dagger$, then

$$AA^\dagger = (AA^\dagger A)A^\dagger = \left(AA^\dagger\right)^2 = \left(AA^\dagger\right)^*.$$

By Lemma 5.3.4, $AA^\dagger \in \mathbb{H}^{m\times m}$. $AA^\dagger$ is hermitian and hence, by Lemma 5.3.6, there exists a unitary matrix $U \in \mathbb{H}^{m\times m}$ such that $U^*AA^\dagger U = D$ where D is diagonal. Since

$$D^2 = (U^*AA^\dagger U)(U^*AA^\dagger U) = U^*AA^\dagger AA^\dagger U = U^*AA^\dagger U = D,$$

the diagonal entries of D are either 1 or 0. Therefore, we can rearrange the rows and columns of U so that $D = \begin{pmatrix} I_r & 0 \\ 0 & 0 \end{pmatrix}$ with $r \le \min\{m, n\}$.

Set $B = U^*A$. By Lemma 5.3.3, B has its own generalized inverse $B^\dagger$ and $BB^\dagger = \begin{pmatrix} I_r & 0 \\ 0 & 0 \end{pmatrix}$. Write B as a blocked matrix form, that is, $B = \begin{pmatrix} A_1 & A_2 \\ A_3 & A_4 \end{pmatrix}$, for some arbitrary quaternion polynomial matrices $A_1 \in \mathbb{H}[x]^{r\times r}$, $A_2 \in \mathbb{H}[x]^{r\times(n-r)}$, $A_3 \in \mathbb{H}[x]^{(m-r)\times r}$, and $A_4 \in \mathbb{H}[x]^{(m-r)\times(n-r)}$. Since $B = BB^\dagger B = \begin{pmatrix} I_r & 0 \\ 0 & 0 \end{pmatrix}\begin{pmatrix} A_1 & A_2 \\ A_3 & A_4 \end{pmatrix} = \begin{pmatrix} A_1 & A_2 \\ 0 & 0 \end{pmatrix}$, we must have $A_3 = 0, A_4 = 0$, and thus $BB^* = \begin{pmatrix} A_1A_1^* + A_2A_2^* & 0 \\ 0 & 0 \end{pmatrix}$. Similarly, $B^\dagger = \begin{pmatrix} B_1 & 0 \\ B_2 & 0 \end{pmatrix}$ for some B_1 and B_2. By Lemma 5.3.5,

$$\text{Image}(BB^*) = \text{Image}(B) = \text{Image}(BB^\dagger) = \text{Image}\begin{pmatrix} I_r & 0 \\ 0 & 0 \end{pmatrix}.$$

This implies the surjectivity of $A_1A_1^* + A_2A_2^*$ on $\mathbb{H}[x]^{r\times 1}$. Therefore, $A_1A_1^* + A_2A_2^*$ is a unit in $\mathbb{H}[x]^{r\times r}$ and

$$A = UB = U\begin{pmatrix} A_1 & A_2 \\ 0 & 0 \end{pmatrix}.$$

Next, we have that:

$$
\begin{aligned}
B^{\dagger} &= B^{\dagger}\left(B^{\dagger}\right)^{*} B^{*} = B^{\dagger}\left(B^{*}\right)^{\dagger} B^{*} = B^{*}\left(BB^{*}\right)^{\dagger} \\
&= \begin{pmatrix} A_1^* & 0 \\ A_2^* & 0 \end{pmatrix} \begin{pmatrix} \left(A_1A_1^* + A_2A_2^*\right)^{-1} & 0 \\ 0 & 0 \end{pmatrix} = \begin{pmatrix} A_1^*\left(A_1A_1^* + A_2A_2^*\right)^{-1} & 0 \\ A_2^*\left(A_1A_1^* + A_2A_2^*\right)^{-1} & 0 \end{pmatrix},
\end{aligned}
$$

which gives

$$
A^{\dagger} = \begin{pmatrix} A_1^*\left(A_1A_1^* + A_2A_2^*\right)^{-1} & 0 \\ A_2^*\left(A_1A_1^* + A_2A_2^*\right)^{-1} & 0 \end{pmatrix} U^*.
$$

($\Longleftarrow$) The converse can be proved by direct computation. □

5.4 Leverrier–Faddeev Algorithm

There are many algorithms for computing the Moore–Penrose inverse. In [26], Faddeev provided an algorithm to compute the characteristic polynomial of an $n \times n$ matrix over a field, which is a modification of a method of Levereier (1840). This algorithm is not computational efficiency. But the proof is constructive in a rather clear way. Now the Leverrier–Faddeev algorithm is one of the classical methods that has been used to compute the Moore–Penrose inverse. We refer the reader to [5, 23, 26, 51, 127] for more details.

For a given quaternion polynomial matrix A, our trick is to use a square matrix AA^* instead of A. First we define the characteristic polynomial for quaternion polynomial matrix A by using AA^*, and prove that the coefficients of this characteristic polynomial are reals. Then we show that Leverrier–Faddeev algorithm works very well for quaternion polynomial matrices.

Lemma 5.4.1. *Let $A \in \mathbb{H}[x]^{m\times n}$. Then the eigenvalues of AA^* are real.*

Proof. Let $B = AA^*$ and $\lambda \in \mathbb{H}$ be an eigenvalue of B with corresponding eigenvector $0 \neq \vec{x} = (x_1 \cdots x_m)^T \in \mathbb{H}[x]^{m\times 1}$ such that $B\vec{x} = \vec{x}\lambda$. Then $\vec{x}^*B\vec{x} = \vec{x}^*\vec{x}\lambda$. Note that $B = B^*$. We have that $\vec{x}^*B\vec{x} = \lambda^*\vec{x}^*\vec{x}$, and thus

$$
\vec{x}^*\vec{x}\lambda = \lambda^*\vec{x}^*\vec{x} = \left(\vec{x}^*\vec{x}\lambda\right)^*,
$$

that is,

$$
\begin{aligned}
(\bar{x}_1, \cdots \bar{x}_m) \begin{pmatrix} x_1 \\ \vdots \\ x_m \end{pmatrix} \lambda = \left(\sum_{i=1}^m \bar{x}_i x_i\right)\lambda \quad &= \left(\left(\sum_{i=1}^m \bar{x}_i x_i\right)\lambda\right)^* \\
&= \lambda^*\left(\sum_{i=1}^m \bar{x}_i x_i\right)^* = \lambda^*\left(\sum_{i=1}^m \bar{x}_i x_i\right).
\end{aligned}
$$

By Lemma 5.3.1, $0 \neq \sum_{i=1}^{m} \bar{x}_i x_i \in \mathbb{R}[x]$. The above equation gives $\lambda = \lambda^*$, which implies $\lambda \in \mathbb{R}$. □

The Cayley–Hamilton theorem for quaternion matrices has been extensively studied. A survey can be found in [141]. For $A \in \mathbb{H}[x]^{m\times n}$, if $A = P + Q\mathbf{j}$ with $P, Q \in \mathbb{C}[x]^{m\times n}$, then the complex adjoint of A is defined as

$$\chi_A = \begin{pmatrix} P & Q \\ -\overline{Q} & \overline{P} \end{pmatrix} \in \mathbb{C}[x]^{2m\times 2n}.$$

Next, we define the characteristic polynomial for a quaternion polynomial matrix.

Definition 5.4.2. For $A \in \mathbb{H}[x]^{m\times n}$, let $B = AA^*$ and χ_B be its complex adjoint. Then $f_B(\lambda) = \det(\lambda I_{2m} - \chi_B)$ is called the characteristic polynomial of A.

Remark 5.4.3. By Lemma 5.4.1, λ can be assumed to be a real indeterminate that enjoys the following: $\lambda = \overline{\lambda}$ and λ commutes element-wise with $\mathbb{H}[x]$.

Theorem 5.4.4. *Let $A \in \mathbb{H}[x]^{m\times n}$ and $B = AA^*$. Then $f_B(\lambda) = g(\lambda)^2$ where $g(\lambda) \in (\mathbb{R}[x])[\lambda]$, that is, a polynomial in one determinate λ over polynomial ring $\mathbb{R}[x]$.*

Proof. We first show that $f_B(\lambda) \in (\mathbb{R}[x])[\lambda]$. Note that $B = AA^*$. We have

$$\det\left((\lambda I_{2m} - \chi_B)^T\right) = \det(\lambda I_{2m} - \chi_B) = \det\left((\lambda I_{2m} - \chi_B)^*\right),$$

and thus

$$\det(\lambda I_{2m} - \chi_B) = \overline{\det(\lambda I_{2m} - \chi_B)} = \overline{\det(\lambda I_{2m} - \chi_B)}.$$

Therefore

$$\det(\lambda I_{2m} - \chi_B) = f_B(\lambda) \in (\mathbb{R}[x])[\lambda]. \tag{5.5}$$

Next, we show that $f_B(\lambda) = g(\lambda)^2$ where $g(\lambda) \in (\mathbb{C}[x])[\lambda]$. Let $B = P + Q\mathbf{j}$. It is easy to check that $P^T = \overline{P}$ and $Q = -Q^T$. Therefore,

$$\chi_B = \begin{pmatrix} P & Q \\ -\overline{Q} & \overline{P} \end{pmatrix} = \begin{pmatrix} P & -Q^T \\ -\overline{Q} & P^T \end{pmatrix} \implies \lambda I_{2m} - \chi_B = \begin{pmatrix} \lambda I_m - P & Q^T \\ \overline{Q} & \lambda I_m - P^T \end{pmatrix}.$$

Next, we have

$$\begin{pmatrix} I_m & -I_m \\ 0 & I_m \end{pmatrix} \begin{pmatrix} I_m & 0 \\ I_m & I_m \end{pmatrix} \begin{pmatrix} I_m & -I_m \\ 0 & I_m \end{pmatrix} \begin{pmatrix} \lambda I_m - P & Q^T \\ \overline{Q} & \lambda I_m - P^T \end{pmatrix} = \begin{pmatrix} \overline{Q} & P^T - \lambda I_m \\ \lambda I_m - P & Q^T \end{pmatrix}.$$

Therefore,

$$f_B(\lambda) = \det\begin{pmatrix} \lambda I_m - P & Q^T \\ \overline{Q} & \lambda I_m - P^T \end{pmatrix} = \det\begin{pmatrix} \overline{Q} & P^T - \lambda I_m \\ \lambda I_m - P & Q^T \end{pmatrix}.$$

Note that

$$\begin{pmatrix} \overline{Q} & P^T - \lambda I_m \\ \lambda I_m - P & Q^T \end{pmatrix}^T = -\begin{pmatrix} \overline{Q} & P^T - \lambda I_m \\ \lambda I_m - P & Q \end{pmatrix},$$

which implies that $\begin{pmatrix} \overline{Q} & P^T - \lambda I_m \\ \lambda I_m - P & Q^T \end{pmatrix}$ is skew-symmetric. By [74], the determinant of $\begin{pmatrix} \overline{Q} & P^T - \lambda I_m \\ \lambda I_m - P & Q^T \end{pmatrix}$, also called its Pfaffian, can be written as the square of a polynomial in its entries. Therefore, $f_B(\lambda) = g(\lambda)^2$ where $g(\lambda) \in (\mathbb{C}[x])[\lambda]$, a polynomial in one determinate λ over the polynomial ring $\mathbb{C}[x]$.

Finally we show that $g(\lambda) \in (\mathbb{R}[x])[\lambda]$. Suppose otherwise. Then $g(\lambda) = a(\lambda) + b(\lambda)\mathbf{i}$ where $a(\lambda)$ and $b(\lambda) \in (\mathbb{R}[x])[\lambda]$ with $b(\lambda) \neq 0$. By (5.5), $g(\lambda)^2 = a(\lambda)^2 - b(\lambda)^2 + 2a(\lambda)b(\lambda)\mathbf{i} \in (\mathbb{R}[x])[\lambda]$. Hence $a(\lambda) = 0$ and $f_B(\lambda) = g(\lambda)^2 = (b(\lambda)\mathbf{i})^2 = -b(\lambda)^2$ where $b(\lambda) \in (\mathbb{R}[x])[\lambda]$. For a fixed $x \in \mathbb{R}$, let $\lambda' \in \mathbb{R}$ be large enough such that $\lambda' I_{2m} - \chi_B \in \mathbb{C}^{2m\times 2m}$ is diagonally dominant with nonnegative diagonal entries and that $(b(x))(\lambda') \neq 0$. Since $\lambda' I_{2m} - \chi_B$ is also hermitian, $\lambda' I_{2m} - \chi_B$ is positive definite [44]. But $\det(\lambda' I_{2m} - \chi_B) = -((b(x))(\lambda'))^2 < 0$, a contradiction. Therefore, $b(x) = 0$ and thus $f_B(\lambda) = g(\lambda)^2$ where $g(\lambda) \in (\mathbb{R}[x])[\lambda]$. □

Corollary 5.4.5. *Let $A \in \mathbb{H}[x]^{m\times n}$, $B = AA^*$, and $f_B(\lambda) = g(\lambda)^2$. Then $g(B) = 0$. $g(\lambda)$ is said to be the generalized characteristic polynomial of A.*

Proof. Note that $g(\lambda) \in (\mathbb{R}[x])[\lambda]$ by Theorem 5.4.4. Then $\chi_{g(B)} = g(\chi_B)$. Next, $f_B(\chi_B) = 0$ by the Cayley–Hamilton theorem for complex polynomial matrices [44]. Therefore $g(\chi_B) = 0$, and $0 = g(\chi_B) = \chi_{g(B)}$, that is, $g(B) = 0$. □

From the definition, it is easy to check the following lemma, which have analogues in the complex case.

Lemma 5.4.6. *Let $A \in \mathbb{H}[x]^{m\times n}$ such that $A^\dagger$ exists. Set $B = AA^*$. Then*

(i) $B^\dagger = (A^*)^\dagger A^\dagger$ *and* $B^\dagger B = AA^\dagger$.
(ii) $B^\dagger B = BB^\dagger$ *and* $(B^\dagger B)^2 = B^\dagger B$.
(iii) $(B^\dagger)^k = (B^k)^\dagger$ *and* $(B^{n-k})^\dagger B^{n-k} = B^\dagger B$, *for any* $k \in \mathbb{N}$.

The following result is well-known for quaternion matrices and it is easy to check that the result also holds for quaternion polynomials.

Lemma 5.4.7. *Let $A \in \mathbb{H}[x]^{m\times n}$, $B \in \mathbb{H}[x]^{p\times q}$, and $C \in \mathbb{H}[x]^{m\times q}$. If $A^\dagger$ and $B^\dagger$ both exist, then the quaternion polynomial matrix equation $AXB = C$ has a solution in $\mathbb{H}[x]^{n\times p}$ if and only if $AA^\dagger CB^\dagger B = C$, in which case the general solution is*

$$X = A^{\dagger}CB^{\dagger} + Y - A^{\dagger}AYBB^{\dagger},$$

where $Y \in \mathbb{H}[x]^{n\times p}$ is arbitrary.

Theorem 5.4.8. *Let $A \in \mathbb{H}[x]^{m\times n}$ such that $A^{\dagger}$ exists and $B = AA^{*}$. Suppose that the generalized characteristic polynomial of A is*

$$g(\lambda) = \lambda^{m} + a_1\lambda^{m-1} + \cdots + a_k\lambda^{m-k} + \cdots + a_{m-1}\lambda + a_m,$$

where $a_i \in \mathbb{R}[x]$. If k is the largest integer such that $a_k \neq 0$, then the Moore–Penrose inverse of A is given by

$$A^{\dagger} = -\frac{1}{a_k}A^{*}\left[B^{k-1} + a_1B^{k-2} + \cdots + a_{k-1}I\right].$$

If $a_i = 0$ for all $1 \le i \le m$, then $A^{\dagger} = 0$.

Proof. The proof is similar to the complex case in [23] by using Corollary 5.4.5, Lemmas 5.4.6, 5.4.7, and 5.3.3. □

From the above theorem, we can find the Moore–Penrose inverse $A^{\dagger}$ of A by computing its generalized characteristic polynomial. Fadeev [27] modified Leverrier's method and gave an algorithm to compute $\{a_i\}$ without computing $g(\lambda)$. Next, we extend this algorithm to quaternion polynomial matrices.

Lemma 5.4.9. *Let $A \in \mathbb{H}[x]^{m\times n}$ such that $A^{\dagger}$ exists and set $B = AA^{*}$. Then for $1 \le k \le m$,*

$$tr\left[\left(B^{k} + a_1B^{k-1} + \cdots + a_{k-1}B\right)\right] = -ka_k,$$

where the a_i's arise from the following generalized characteristic polynomial of A:

$$g(\lambda) = \lambda^{m} + a_1\lambda^{m-1} + \cdots + a_k\lambda^{m-k} + \cdots + a_{m-1}\lambda + a_m \in (\mathbb{R}(\lambda))[x].$$

Proof. Let $Y = yI$ where $y \in \mathbb{R}$. We can write $g(Y)$ as:

$$\begin{aligned} g(Y) &= g(Y) - g(B) \\ &= (Y - B)\left[Y^{m-1} + (B + a_1I)Y^{m-2} + \cdots + \left(B^{m-1} + a_1B^{m-2} + \cdots + a_mI\right)\right]. \end{aligned}$$

As long as y is not an eigenvalue of B, $yI - B = Y - B$ is nonsingular, so we can write

$$\begin{aligned} (Y - B)^{-1}g(Y) = &Y^{m-1} + (B + a_1I)Y^{m-2} + \left(B^{2} + a_1B + a_2I\right)Y^{m-3} + \cdots \\ &+ \left(B^{m-1} + a_1B^{m-2} + \cdots + a_mI\right). \end{aligned}$$

Taking the traces gives

$$\begin{aligned}&\operatorname{tr}\left[(Y-B)^{-1}g(Y)\right]\\&=my^{m-1}+\operatorname{tr}\left[(B+a_1I)\right]y^{m-2}+\cdots+\operatorname{tr}\left(B^{m-1}+a_1B^{m-2}+\cdots+a_mI\right).\end{aligned}$$

Let $C = (Y-B)^{-1}g(Y)$. Since $g(Y) = g(yI) = g(y)I$, we have $C = g(y)(Y-B)^{-1}$. Therefore,

$$\operatorname{tr}(C) = g(y)\operatorname{tr}\left[(Y-B)^{-1}\right].$$

Let $\lambda_1, \cdots, \lambda_{m'}$, where $m' \le m$, be all the nonzero eigenvalues of B. $\operatorname{tr}\left[(Y-B)^{-1}\right]$ is the sum of the eigenvalues of $(Y-B)^{-1}$. We will show that these eigenvalues are the fractions $\frac{1}{y-\lambda_1}, \cdots, \frac{1}{y-\lambda_{m'}}$.

Let ζ be an eigenvalue of $(Y-B)^{-1}$ with corresponding eigenvector $\vec{v}$ such that:

$$(Y-B)^{-1}\vec{v} = \vec{v}\zeta.$$

ζ is real by Lemma 5.4.1, and hence

$$(Y-B)\vec{v} = \vec{v}\frac{1}{\zeta} \implies B\vec{v} = \vec{v}\left(y-\frac{1}{\zeta}\right).$$

Therefore, $y-\frac{1}{\zeta} = \lambda_i$ implies $\zeta = \frac{1}{y-\lambda_i}$ for some $1 \le i \le m'$.

Since $g(y) = (y-\lambda_1)(y-\lambda_2)\cdots(y-\lambda_{m'})$, we have that the first derivative $g'(y) = g(y)\left(\frac{1}{y-\lambda_1}+\cdots+\frac{1}{y-\lambda_{m'}}\right)$ and $\operatorname{tr}(C) = g'(y)$. On the other hand, the derivative of g is also equal to:

$$g'(y) = my^{m-1} + a_1(m-1)y^{m-2} + \cdots + a_{m-1}.$$

Therefore,

$$\begin{aligned}&my^{m-1}+a_1(m-1)y^{m-2}+\cdots+a_{m-1}\\&=my^{m-1}+\operatorname{tr}(B+a_1I)y^{m-2}+\cdots+\operatorname{tr}\left(B^{m-1}+a_1B^{m-2}+\cdots+a_mI\right).\end{aligned}$$

Comparing the coefficient of y^{m-k-1} on both sides, we obtain

$$\begin{aligned}a_k(m-k) &= \operatorname{tr}\left(B^k+a_1B^{k-1}+\cdots+a_{k-1}B+a_kI\right)\\&= \operatorname{tr}\left(B^k+a_1B^{k-1}+\cdots+a_{k-1}B\right)+\operatorname{tr}(a_kI),\end{aligned}$$

and so

$$-ka_k = \operatorname{tr}\left(B^k + a_1 B^{k-1} + \cdots + a_{k-1} B\right).$$

□

Now the question is changed to find the coefficients of the generalized characteristic polynomial in order to compute the Moore–Penrose inverse. Next, we present the Leverrier–Faddeev algorithm for finding Moore–Penrose inverses of quaternion polynomial matrices by recursively computing traces.

Proposition 5.4.10. *Let $A \in \mathbb{H}[x]^{m\times n}$ such that $A^\dagger$ exists and $B = AA^*$. Suppose that the generalized characteristic polynomial of A is*

$$g(\lambda) = \lambda^m + a_1\lambda^{m-1} + \cdots + a_k\lambda^{m-k} + \cdots + a_{m-1}\lambda + a_m,$$

where $a_i \in \mathbb{R}[x]$. Define $a_0 = 1$. If p is the largest integer such that $a_p \neq 0$ and we construct the sequence $A_0, \cdots, A_p$ as follows:

$$\begin{array}{llll}
A_0 & = 0 & -1 = q_0 & B_0 = I \\
A_1 & = AA^*B_0 & \frac{trA_1}{1} = q_1 & B_1 = A_1 - q_1 I \\
\vdots & & \vdots & \vdots \\
A_{p-1} & = AA^*B_{p-2} & \frac{trA_{p-1}}{p-1} = q_{p-1} & B_{p-1} = A_{p-1} - q_{p-1}I \\
A_p & = AA^*B_{p-1} & \frac{trA_p}{p} = q_p & B_p = A_p - q_p I
\end{array}$$

then $q_i(x) = -a_i(x)$, $i = 0, \cdots, p$.

Proof. We will show $q_i(x) = -a_i(x)$ by mathematical induction. By the definition, clearly $q_0 = -a_0$ holds.

Now we assume that $q_i(x) = -a_i(x)$ holds for all $0 \le i \le k-1$. Then

$$\begin{aligned}
A_k &= AA^*B_{k-1} \\
&= BB_{k-1} \\
&= B(A_{k-1} - q_{k-1}I) \\
&= B(B(A_{k-2} - q_{k-2}I) - q_{k-1}I) \\
&\cdots \\
&= B^k - q_1B^{k-1} - q_2B^{k-2} - \cdots - q_{k-1}B \\
&= B^k + a_1B^{k-1} + a_2B^{k-2} + \cdots + a_{k-1}B.
\end{aligned}$$

and thus

$$\operatorname{tr}(A_k) = \operatorname{tr}\left(B^k + a_1B^{k-1} + \cdots + a_{k-1}B\right),$$

which, by Lemma 5.4.9, is equal to $-ka_k$. So $q_k = \frac{\operatorname{tr} A_k}{k} = -a_k$. Therefore, $q_i(x) = -a_i(x)$ for all $p \geq i \geq 0$. □

Now, combining Theorem 5.4.8 and Proposition 5.4.10, we have the following algorithm to compute the Moore–Penrose inverse:

Algorithm 3 Leverrier–Faddeev algorithm for quaternion polynomial matrices

---

Input $A \in \mathbb{H}[x]^{m \times n}$
Output The Moore–Penrose inverse $A^\dagger$ of A in $\mathbb{H}[x]^{n \times m}$ if exists
1: $B_0 \leftarrow I_m$, $a_0 \leftarrow 1$
2: for $i = 1, \ldots, m$ do
$A_i \leftarrow AA^*B_{i-1}$, $a_i \leftarrow -\frac{\operatorname{tr} A_i}{i}$, $B_i \leftarrow A_i + a_i I_m$
3: Find the maximal index p such that $a_p \neq 0$.
4: Return $A^\dagger = \begin{cases} -\frac{1}{a_p} A^* B_{p-1}, & p > 0, \\ 0, & p = 0. \end{cases}$

Note that we have to compute many matrix products in Algorithm 3, which means that Leverrier–Faddeev method is not efficient. In the next section, we will present a more efficient way by combining Theorem 5.4.8 and interpolation methods.

5.5 Finding Moore–Penrose Inverses by Interpolation

Interpolation is an efficient method in many computational questions over commutative fields. In non-commutative case like $\mathbb{H}$, the situation becomes very complicated since some basic properties fail, for example, for a given quaternion polynomial, there might have infinite roots and infinite factors. To overcome this difficulty, we choose the interpolation at data points of real numbers and present an efficient method to obtain the Moore–Penrose inverse of a quaternion polynomial matrix.

Recently, there are few papers regarding non-commutative interpolations and applications (see, for example, [46, 67, 143]). Lets recall some important concepts and properties. An element $r \in \mathbb{H}$ is a root of a nonzero polynomial $f = a_n x^n + \cdots + a_0 \in \mathbb{H}[x]$ if $a_n r^n + \cdots + a_0 = 0$. Since $\mathbb{H}$ is a principal idea domain, using Euclidean Algorithm, it is easy to see that $f(r) = 0$ if and only if $x - r$ is a right divisor of f. The set of polynomials in $\mathbb{H}[x]$ having r as a root is the left ideal $\mathbb{H}[x] \cdot (x - r)$. It is worth mentioning that the evaluations of quaternion polynomials are quite different from the commutative case. It is defined as following: let f, g and $h \in \mathbb{H}[x]$, $f = gh$ and $r \in \mathbb{H}$. If $h(r) = 0$, then $f(r) = 0$. Otherwise, set $\beta = h(r) \neq 0$. Then the evaluation of $f(x)$ at $x = r$ is

$$f(r) = g\left(\beta r \beta^{-1}\right) h(r). \tag{5.6}$$

In particular, if r is a root of f but not of h, then $\beta r \beta^{-1}$ is a root of g. We refer the reader to [63] for more details.

Although a quaternion could have infinite many roots, in [38], it is proved that if $f \in \mathbb{H}[x]$ is of degree n, then the roots of f lie in at most n conjugacy classes of $\mathbb{H}$.

It is well-known that Newton's interpolation and Lagrange's interpolation play important roles in studying polynomials over fields. Unfortunately, one cannot get similar nice formulas in quaternion case. Fortunately, we can still compute a quaternion polynomial from a given set of pairs of quaternions.

Lemma 5.5.1. *Let $c_1, \ldots, c_n$ be n pairwise non-conjugate elements of $\mathbb{H}$. Then there is a unique monic polynomial $g_n \in \mathbb{H}[x]$ of degree n such that $g_n(c_1) = \cdots = g_n(c_n) = 0$. Moreover, $c_1, \ldots, c_n$ are the only roots (up to conjugacy classes) of g_n in $\mathbb{H}$.*

Proof. We first show the existence of g_n for all $n \geq 1$ by mathematical induction. For $n = 1$, it is trivially true as $g_1 = x - c_1$.

Suppose the claim holds for all $1 \leq n \leq k-1$. Let $c_1, \cdots, c_k \in \mathbb{H}$ be pairwise non-conjugate. Invoking the inductive hypothesis, there exists a monic polynomial g_{k-1} of degree $k-1$ with $c_2, \cdots, c_k$ as its only roots (up to conjugacy classes), that is, $g_{k-1}(c_1) \neq 0$. Construct g_k as follows:

$$g_k(x) = \left(x - g_{k-1}(c_1)\, c_1 g_{k-1}(c_1)^{-1}\right) \cdot g_{k-1}(x).$$

By Eq. (5.6), $g_k(c_1) = 0$. Thus, the claim holds for k. Therefore this claim holds for all $n \geq 1$.

We next show that g_n is unique. For a fixed n, let $g \neq g_n$ be a monic polynomial of degree n such that $g(c_1) = \cdots = g(c_n) = 0$, too. Then $\deg(g_n - g) \leq n-1$ but $g_n - g$ has roots $c_1, \ldots, c_n$ which lie in n different conjugacy classes of $\mathbb{H}$, a contradiction. Therefore, g_n is unique for all $n \geq 1$. □

Proposition 5.5.2. *Let $c_1, \ldots, c_{n+1} \in \mathbb{H}$ be pairwise non-conjugate and let $d_1, \ldots, d_{n+1} \in \mathbb{H}$. Then there exists a unique lowest degree polynomial $f \in \mathbb{H}[x]$, of degree $p \leq n$, such that $f(c_i) = d_i$ for all $1 \leq i \leq n+1$.*

Proof. For any $1 \leq s \leq n+1$, let $S = \{1, \cdots, n+1\} \setminus \{s\}$. By Lemma 5.5.1, we can find a unique monic $h_S \in \mathbb{H}[x]$ of degree n such that $h_S(c_i) = 0,\ i \in S$ and that $\{c_i \mid i \in S\}$ are the only roots (up to conjugacy class) of h_S in $\mathbb{H}$. Then $h_S(c_s) \neq 0$, and thus we can construct a quaternion polynomial g_S of degree n such that

$$g_S(c_\alpha) = \begin{cases} 0 & \alpha \in S, \\ 1 & \alpha = s, \end{cases}$$

as follows:

$$g_S(x) = h_S(c_s)^{-1}\, h_S(x).$$

Furthermore, we construct a quaternion polynomial f of degree at most n such that $f(c_i) = d_i$ for all $1 \leq i \leq n+1$ as follows:

$$f = \sum_{s=1}^{n+1} d_s g_S.$$

Finally, we show that f is unique. Suppose we have $f_1 \in \mathbb{H}[x]$ of degree $p_1 \leq n$ such that $f_1 \neq f$ and that $f_1(c_i) = d_i$ for all $1 \leq i \leq n+1$, too. Then $f - f_1 \neq 0$ is of degree at most n. But $f - f_1$ has roots $c_1, \ldots, c_{n+1}$ which lie in $n+1$ conjugacy classes of $\mathbb{H}$, a contradiction. Therefore, f is unique. □

From above proof, we can see that it is impossible to construct the so-called Newton divided difference formula for quaternion polynomials. Next we extend the interpolation to quaternion polynomial matrices. Recall that the degree of a given $A \in \mathbb{H}[x]^{m\times n}$ is defined as

$$\deg A = \max\left\{\deg\left(A_{ij}\right) \mid 1 \leq i \leq m, 1 \leq j \leq n\right\}.$$

The following lemma estimates the upper bound of the degree of its Moore–Penrose inverse $A^{\dagger}$ (if it exists).

Lemma 5.5.3. *Let $A \in \mathbb{H}[x]^{m\times n}$ such that $A^{\dagger}$ exists. Then*

$$\deg A^{\dagger} \leq (2m-1)\deg A.$$

Proof. By Theorem 5.4.8,

$$\deg A^{\dagger} \leq \deg\left(A^{*}\left(AA^{*}\right)^{m-1}\right) \leq \deg\left(A^{2m-1}\right) \leq (2m-1)\deg A.$$

□

For $A = (A_{ij}) \in \mathbb{H}[x]$ and $c \in \mathbb{H}$, the evaluation of A at c can be defined as entrywise in a common sense, that is, $A(c) = (A_{ij}(c))$. One has to pay an attention that the evaluations of quaternion polynomials have some special rules as we explained at the beginning of this section.

Proposition 5.5.4. *Let $c_1, \cdots, c_{k+1} \in \mathbb{H}$ be pairwise non-conjugate and let $A_1, \cdots, A_{k+1} \in \mathbb{H}^{n\times m}$. Then there is a unique lowest degree matrix $A \in \mathbb{H}[x]^{n\times m}$ of degree $p \leq k$, such that $A(c_i) = A_i$ for all $1 \leq i \leq k+1$.*

Proof. For any $1 \leq n_1 \leq n$ and $1 \leq m_1 \leq m$, by Proposition 5.5.2, there is a lowest degree polynomial $A_{n_1m_1}(x)$ determined by the values $c_1, \cdots, c_{k+1}$ and $(A_1)_{n_1m_1}, \ldots, (A_{k+1})_{n_1m_1}$. In fact, for any $1 \leq s \leq k+1$, let $S = \{1, \cdots, k+1\} \setminus \{s\}$. Then

$$A_{n_1m_1}(x) = \sum_{s=1}^{k+1} (A_s)_{n_1m_1} g_S(x),$$

where $g_S(c_\alpha) = \begin{cases} 0 & \alpha \in S, \\ 1 & \alpha = s \end{cases}$. Since n_1 and m_1 are chosen randomly, the lowest degree matrix A that satisfies $A(c_i) = A_i$ for all $1 \le i \le k+1$ is determined by $A = (\sum_{s=1}^{k+1} A_s g_S)$.

Next we show that A is unique. Suppose $C \neq A$ of degree $p' \le p$ also satisfies $C(c_i) = A_i$ for all $1 \le i \le k+1$. Then for some $1 \le n_2 \le n$ and $1 \le m_2 \le m$, $(A - C)_{n_2 m_2} \neq 0$. But $(A - C)_{n_2 m_2}$, of degree at most $p \le k$, has roots $c_1, \ldots, c_{k+1}$ which lie in $k+1$ conjugacy classes of $\mathbb{H}$, a contradiction. Therefore, A is unique. □

Let $A \in \mathbb{H}[x]^{m \times n}$ such that $A^\dagger$ exists, and set $B = AA^*$. Let p be the largest integer such that $a_p \neq 0$. We can construct the sequence $A_0, \ldots, A_p$ as in Proposition 5.4.10. The next theorem presents the interpolation version of Leverrier–Faddev algorithm.

Theorem 5.5.5. *In the above setting, let $k = (2m-1)\deg A$ and $c_1, \ldots, c_{k+1} \in \mathbb{R}$ be $k+1$ distinct real numbers such that $q_p(c_s) \neq 0$ for any $1 \le s \le k+1$. Let $S = \{1, \cdots, k+1\} \setminus \{s\}$. Then*

$$A^\dagger = \sum_{s=1}^{k+1} A(c_s)^\dagger g_S$$

where

$$A(c_s)^\dagger = \frac{1}{q_p(c_s)} A(c_s)^* \left[B(c_s)^{p-1} - q_1(c_s) B(c_s)^{p-2} - \cdots - q_{p-1}(c_s) I \right]$$

and

$$g_S(c_\alpha) = \begin{cases} 0 & \alpha \in S, \\ 1 & \alpha = s. \end{cases}$$

Proof. It follows from Theorem 5.4.8, Propositions 5.4.10 and 5.5.4. □

The upper bound of degrees of $A^\dagger$ in Lemma 5.5.3 is not sharp. In fact, in many questions, one only needs to pick up a few real points. (see Example 5.6.1)

5.6 Implementations and Examples

The calculations of quaternions are very complicated and time-consuming. It is almost impossible to do some calculations for quaternion polynomials and quaternion polynomial matrices even for a small sized matrices by hand. There are only few quaternion packages in the computer algebra system Maple. But none of these has commands for quaternion polynomials and quaternion polynomial matrices. In [47], we developed a Maple package which includes all basic operations for quaternion polynomials and quaternion polynomial matrices. In particular, all the algorithms in this chapter were implemented. We give the following illustrative example:

Example 5.6.1. Let us consider the problem of determining the Moore–Penrose inverse of the following quaternion polynomial matrix:

$$A = \begin{pmatrix} 14x+14+76\mathbf{i}+70\mathbf{j}+56\mathbf{k} & 56-28\mathbf{i}-70\mathbf{j}+70\mathbf{k} & 28\mathbf{j}-56\mathbf{k} & 14x-56-8\mathbf{i}-14\mathbf{j}-56\mathbf{k} \\ -2x-2-43\mathbf{i}-10\mathbf{j}-8\mathbf{k} & -8+4\mathbf{i}+10\mathbf{j}-10\mathbf{k} & -4\mathbf{j}+8\mathbf{k} & -2x+8-31\mathbf{i}+2\mathbf{j}+8\mathbf{k} \\ -3x-3+3\mathbf{i}-15\mathbf{j}-12\mathbf{k} & -12+6\mathbf{i}+15\mathbf{j}-15\mathbf{k} & -6\mathbf{j}+12\mathbf{k} & -3x+12+21\mathbf{i}+3\mathbf{j}+12\mathbf{k} \\ -4x-4+4\mathbf{i}-20\mathbf{j}-16\mathbf{k} & -16+8\mathbf{i}+20\mathbf{j}-20\mathbf{k} & -8\mathbf{j}+16\mathbf{k} & -4x+16+28\mathbf{i}+4\mathbf{j}+16\mathbf{k} \end{pmatrix} \in \mathbb{H}^{4\times 4}[x].$$

From Lemma 5.5.3, we know that the upper bound of the degree of $A^{\dagger}$ is less than $(2m-1)\deg A = (2\times 4-1)\cdot 1 = 7$. In practice, we don't need to start from the upper bound. For this example, we may guess $\deg A^{\dagger} = 2$, and choose $c_1 = 0$ and $c_2 = 1$. Then using our Maple package, it is easy to do the following calculations:

$$A(c_1) = \begin{pmatrix} 14+76\mathbf{i}+70\mathbf{j}+56\mathbf{k} & 56-28\mathbf{i}-70\mathbf{j}+70\mathbf{k} & 28\mathbf{j}-56\mathbf{k} & -56-8\mathbf{i}-14\mathbf{j}-56\mathbf{k} \\ -2-43\mathbf{i}-10\mathbf{j}-8\mathbf{k} & -8+4\mathbf{i}+10\mathbf{j}-10\mathbf{k} & -4\mathbf{j}+8\mathbf{k} & 8-31\mathbf{i}+2\mathbf{j}+8\mathbf{k} \\ -3+3\mathbf{i}-15\mathbf{j}-12\mathbf{k} & -12+6\mathbf{i}+15\mathbf{j}-15\mathbf{k} & -6\mathbf{j}+12\mathbf{k} & 12+21\mathbf{i}+3\mathbf{j}+12\mathbf{k} \\ -4+4\mathbf{i}-20\mathbf{j}-16\mathbf{k} & -16+8\mathbf{i}+20\mathbf{j}-20\mathbf{k} & -8\mathbf{j}+16\mathbf{k} & 16+28\mathbf{i}+4\mathbf{j}+16\mathbf{k} \end{pmatrix}$$

and

$$A(c_2) = \begin{pmatrix} 28+76\mathbf{i}+70\mathbf{j}+56\mathbf{k} & 56-28\mathbf{i}-70\mathbf{j}+70\mathbf{k} & 28\mathbf{j}-56\mathbf{k} & -42-8\mathbf{i}-14\mathbf{j}-56\mathbf{k} \\ -4-43\mathbf{i}-10\mathbf{j}-8\mathbf{k} & -8+4\mathbf{i}+10\mathbf{j}-10\mathbf{k} & -4\mathbf{j}+8\mathbf{k} & 6-31\mathbf{i}+2\mathbf{j}+8\mathbf{k} \\ -6+3\mathbf{i}-15\mathbf{j}-12\mathbf{k} & -12+6\mathbf{i}+15\mathbf{j}-15\mathbf{k} & -6\mathbf{j}+12\mathbf{k} & 9+21\mathbf{i}+3\mathbf{j}+12\mathbf{k} \\ -8+4\mathbf{i}-20\mathbf{j}-16\mathbf{k} & -16+8\mathbf{i}+20\mathbf{j}-20\mathbf{k} & -8\mathbf{j}+16\mathbf{k} & 12+28\mathbf{i}+4\mathbf{j}+16\mathbf{k} \end{pmatrix}.$$

By the algorithm stated in Theorem 5.5.5, we calculate and obtain

$$A(c_1)^{\dagger} = A(0)^{\dagger} = \frac{1}{230175}\times$$
$$\begin{pmatrix} 140-560\mathbf{i}-228\mathbf{j}-342\mathbf{k} & 355+1730\mathbf{i}-96\mathbf{j}+81\mathbf{k} & -255-870\mathbf{i}+126\mathbf{j}+54\mathbf{k} & -340-1160\mathbf{i}+168\mathbf{j}+72\mathbf{k} \\ 276+88\mathbf{i}+426\mathbf{j}-382\mathbf{k} & 282+416\mathbf{i}-93\mathbf{j}-149\mathbf{k} & -252-276\mathbf{i}-72\mathbf{j}+204\mathbf{k} & -336-368\mathbf{i}-96\mathbf{j}+272\mathbf{k} \\ 32+16\mathbf{i}-176\mathbf{j}+292\mathbf{k} & -176-88\mathbf{i}+68\mathbf{j}+194\mathbf{k} & 96+48\mathbf{i}+12\mathbf{j}-204\mathbf{k} & 128+64\mathbf{i}+16\mathbf{j}-272\mathbf{k} \\ -140-122\mathbf{i}+228\mathbf{j}+342\mathbf{k} & -355+2021\mathbf{i}+96\mathbf{j}-81\mathbf{k} & 255-1176\mathbf{i}-126\mathbf{j}-54\mathbf{k} & 340-1568\mathbf{i}-168\mathbf{j}-72\mathbf{k} \end{pmatrix}$$

and

$$A(c_2)^{\dagger} = A(1)^{\dagger} = \frac{1}{230175}\times$$
$$\begin{pmatrix} 152-550\mathbf{i}-244\mathbf{j}-330\mathbf{k} & 289+1675\mathbf{i}-8\mathbf{j}+15\mathbf{k} & -219-840\mathbf{i}+78\mathbf{j}+90\mathbf{k} & -292-1120\mathbf{i}+104\mathbf{j}+120\mathbf{k} \\ 268+104\mathbf{i}+406\mathbf{j}-402\mathbf{k} & 326+328\mathbf{i}+17\mathbf{j}-39\mathbf{k} & -276-228\mathbf{i}-132\mathbf{j}+144\mathbf{k} & -368-304\mathbf{i}-176\mathbf{j}+192\mathbf{k} \\ 32+16\mathbf{i}-160\mathbf{j}+300\mathbf{k} & -176-88\mathbf{i}-20\mathbf{j}+150\mathbf{k} & 96+48\mathbf{i}+60\mathbf{j}-180\mathbf{k} & 128+64\mathbf{i}+80\mathbf{j}-240\mathbf{k} \\ -152-132\mathbf{i}+244\mathbf{j}+330\mathbf{k} & -289+2076\mathbf{i}+8\mathbf{j}-15\mathbf{k} & 219-1206\mathbf{i}-78\mathbf{j}-90\mathbf{k} & 292-1608\mathbf{i}-104\mathbf{j}-120\mathbf{k} \end{pmatrix}.$$

By Theorem 5.5.5, we have

$$\sum_{s=1}^{2} A(c_s)^{\dagger} g_S = A(0)^{\dagger}(1-x) + A(1)^{\dagger} x$$

$$= \frac{1}{230175} \times$$

$$\left(\begin{matrix} (12+10\mathbf{i}-16\mathbf{j}+12\mathbf{k})x+140-560\mathbf{i}-228\mathbf{j}-342\mathbf{k} & (-66-55\mathbf{i}+88\mathbf{j}-66\mathbf{k})x+355+1730\mathbf{i}-96\mathbf{j}+81\mathbf{k} \\ (-8+16\mathbf{i}-20\mathbf{j}-20\mathbf{k})x+276+88\mathbf{i}+426\mathbf{j}-382\mathbf{k} & (44-88\mathbf{i}+110\mathbf{j}+110\mathbf{k})x+282+416\mathbf{i}-93\mathbf{j}-149\mathbf{k} \\ (16\mathbf{j}+8\mathbf{k})x+32+16\mathbf{i}-176\mathbf{j}+292\mathbf{k} & (-88\mathbf{j}-44\mathbf{k})x-176-88\mathbf{i}+68\mathbf{j}+194\mathbf{k} \\ (-12-10\mathbf{i}+16\mathbf{j}-12\mathbf{k})x-140-122\mathbf{i}+228\mathbf{j}-342\mathbf{k} & (66+55\mathbf{i}-88\mathbf{j}+66\mathbf{k})x-355+2021\mathbf{i}+96\mathbf{j}-81\mathbf{k} \end{matrix}\right.$$

$$\left.\begin{matrix} (36+30\mathbf{i}-48\mathbf{j}+36\mathbf{k})x-255-870\mathbf{i}+126\mathbf{j}+54\mathbf{k} & (48+40\mathbf{i}-64\mathbf{j}+48\mathbf{k})x-340-1160\mathbf{i}+168\mathbf{j}+72\mathbf{k} \\ (-24+48\mathbf{i}-60\mathbf{j}-60\mathbf{k})x-252-276\mathbf{i}-72\mathbf{j}+204\mathbf{k} & (-32+64\mathbf{i}-80\mathbf{j}+80\mathbf{k})x-366-368\mathbf{i}-96\mathbf{j}+272\mathbf{k} \\ (48\mathbf{j}+24\mathbf{k})x+96+48\mathbf{i}+12\mathbf{j}-204\mathbf{k} & (64\mathbf{j}+32\mathbf{k})x+128+64\mathbf{i}+16\mathbf{j}-272\mathbf{k} \\ (-36-30\mathbf{i}+48\mathbf{j}-36\mathbf{k})x+255-1176\mathbf{i}-126\mathbf{j}-54\mathbf{k} & (-48-40\mathbf{i}+64\mathbf{j}-48\mathbf{k})x+340-1568\mathbf{i}-168\mathbf{j}-72\mathbf{k} \end{matrix}\right).$$

It is easy to verify that $\sum_{s=1}^{2} A(c_s)^{\dagger} g_S$ satisfies the four defining relations of the Moore–Penrose inverse. Therefore it is the Moore–Penrose inverse of A.

Chapter 6
M-Matrices over Infinite Dimensional Spaces

6.1 Introduction

The intention here is to present an overview of some very recent results on three classes of operators, extending the corresponding matrix results. The relevant notions that are generalized here are that of a P-matrix, a Q-matrix, and an M-matrix. It is widely known (in the matrix case) that these notions coincide for Z-matrices. While we are not able to prove such a relationship between these classes of operators over Hilbert spaces, nevertheless, we are able to establish a relationship between Q-operators and M-operators, extending an analogous matrix result. It should be pointed out that, in any case, for P-operators, some interesting generalizations of results for P-matrices vis-a-vis invertibility of certain intervals of matrices have been obtained. These were proved by Rajesh Kannan and Sivakumar [92]. Since these are new, we include proofs for some of the important results. The last section considers a class of operators that are more general than M-operators. In particular, we review results relating to the nonnegativity of the Moore–Penrose inverse of Gram operators over Hilbert spaces, reporting the work of Kurmayya and Sivakumar [61] and Sivakumar [125]. These results find a place here is due to the reason that they extend the applicability of results for certain subclasses of M-matrices to infinite dimensional spaces.

A real square matrix A is called a *P-matrix* if all its principal minors are positive. Such a matrix can be characterized by what is known as the *sign non-reversal property*. Taking cue from this, the authors of [92] introduced the notion of a P-operator for infinite dimensional spaces as the first objective. Relationships between invertibility of some subsets of intervals of operators and certain P-operators were then established. These generalize the corresponding results in the matrix case. The inheritance of the property of a P-operator by the Schur complement and the principal pivot transform was also proved. If A is an invertible M-matrix, then there is a positive vector whose image under A is also positive. As the second goal, this and another result on intervals of M-matrices were generalized to operators

P.N. Shivakumar et al., *Infinite Matrices and Their Recent Applications*,
DOI 10.1007/978-3-319-30180-8_6

over Banach spaces. Towards the third objective, the concept of a Q-operator is proposed, generalizing the well-known Q-matrix property. An important result, which establishes connections between Q-operators and invertible M-operators, is proved for Hilbert space operators.

6.2 Preliminary Notions

In this section, we recall some preliminary notions and results that will be referred to in the sequel. We also make a short survey of the literature on relevant results.

We shall be interested in certain matrix classes. Let us start with the first one. As mentioned in the introduction, a real square matrix is called a P-matrix if all its principal minors are positive. It is well known that $A \in \mathbb{R}^{n\times n}$ is a P-matrix if and only if A does not reverse the sign of any nonzero vector, viz.,

$$x_i(Ax)_i \leq 0,\ i = 1, 2, \ldots, n \Longrightarrow x = 0.$$

Fiedler and Ptak [34] proved this result and also established other results for P-matrices, especially in the class of Z-matrices. The first objective of this article is to introduce the notion of P-operators for infinite dimensional Banach spaces with a Schauder basis. We generalize well-known relationships between P-matrices and certain specific subsets of intervals of matrices to intervals of operators over certain Banach spaces. These are presented in Sect. 6.3.

We will be dealing with the notion of the linear complementarity problem. Let us recall that, for $A \in \mathbb{R}^{n\times n}$ and $q \in \mathbb{R}^n$, the *linear complementarity problem* written as $LCP(A, q)$ is to find a vector $x \in \mathbb{R}^n$ such that

$$x \geq 0,\ \ Ax + q \geq 0 \ \text{ and } \ \langle x, Ax + q\rangle = 0.$$

In the context of the linear complementarity problem, it is a well-known fact that A is a P-matrix if and only if $LCP(A, q)$ has a unique solution for all $q \in \mathbb{R}^n$. Another class is defined next. $A \in \mathbb{R}^{n\times n}$ is called a *Q-matrix* if $LCP(A, q)$ has a solution for all $q \in \mathbb{R}^n$. Using the characterization mentioned just now, it follows that a P-matrix must be a Q-matrix, but the converse is not true.

A frequently used result for identifying Q-matrices is the famous Karamardian theorem (Theorem 7.3.1), [56]. We state this result next. Let K be a solid and closed cone in $\mathbb{R}^n$, and K^* its dual cone (the precise definitions are given below). Let $A \in \mathbb{R}^{n\times n}$. If the complementarity problem $LCP(A, q)$ has a unique solution (namely, zero) for $q = 0$, and for some $q \in int(K^*)$, then it has solution for all $q \in \mathbb{R}^n$.

The third class of matrices are M-matrices. This notion was discussed briefly in Chap. 2, Sect. 2.1. For M-matrices, the following result is widely known:

Theorem 6.2.1 (Theorem 2.3, Chap. 6, [11]). *For a Z-matrix $A \in \mathbb{R}^{n\times n}$, the following statements are equivalent:*

(a) $A = sI - B$, for some $B \geq 0$ with $s > \rho(B)$.
(b) There exists $x > 0$ such that $Ax > 0$.
(c) A^{-1} exists and $A^{-1} \geq 0$.

Let us recall that the condition $A^{-1} \geq 0$ was referred to as *inverse positivity* in Sect. 2.1. We refer the reader to the excellent books by Berman and Plemmons [11] and by Cottle et al. [21] for more details and the relationships between the various matrix classes in the context of the linear complementarity problem.

Let us recall another result for M-matrices, this time, in connection with matrix intervals. For $A, B \in \mathbb{R}^{m\times n}$ with $A \leq B$, let $J(A,B) := [A,B]$ be the set of all matrices C such that $A \leq C \leq B$, where we denote $X \geq Y$ if $X - Y \geq 0$. Let us now state a particular case of a result of Rohn (Theorem 1) [97]. Let $A, B \in \mathbb{R}^{n\times n}$ with $A \leq B$. Then A and B are invertible M-matrices if and only if every $C \in J(A,B)$ is an invertible M-matrix.

In what follows, we present a brief survey of some extensions of M-matrices. First, we consider the finite dimensional case. For the past decade and a half, generalizations of M-matrices have been studied quite extensively in certain non-associative objects called *Euclidean Jordan algebras*. We just mention that the real Euclidean space and the space of all real symmetric matrices are examples of such an algebraic structure. In this connection, an interesting result was proved by Gowda and Tao [40] (Theorem 7), which characterizes M-matrices, in terms of Q-matrices. The main result of this chapter, viz., Theorem 6.5.1, is motivated by this result.

In order to study extensions of the results discussed above, we recall the notion of cones in vector spaces. Let X be a real linear space. Then X is called a *partially ordered vector space* if there is a partial order "$\leq$" defined on X such that the following compatibility conditions are satisfied:

$$x \leq y \Longrightarrow x + z \leq y + z \text{ for all } z \in X$$

and

$$x \leq y \Longrightarrow \alpha x \leq \alpha y \text{ for all } \alpha \geq 0.$$

A subset X_+ of a real linear space X is said to be a *cone* if, $X_+ + X_+ \subseteq X_+$, $\alpha X_+ \subseteq X_+$ for all $\alpha \geq 0$, $X_+ \cap -X_+ = \{0\}$ and $X_+ \neq \{0\}$. A vector $x \in X$ is said to be *nonnegative*, if $x \in X_+$. This is denoted by $x \geq 0$. We define $x \leq y$ if and only if $y - x \in X_+$. Then "$\leq$" is a partial order (induced by X_+) on X. Conversely, if X is a partially ordered normed linear space with the partial order "$\leq$", then the set

$$X_+ = \{x \in X : x \geq 0\}$$

is a cone, called the *positive cone* of X. By a *partial ordered real normed linear space* X we mean a real normed linear space X together with a (topologically) *closed* positive cone X_+. Let X' denote the space of all continuous linear functionals on X. The dual cone X_+^* of X_+ is defined as follows:

$$X_+^* = \{f \in X' : f(x) \geq 0 \text{ for all } x \in X_+\}.$$

A partially ordered real normed linear space which is also a Banach space is called a *partially ordered Banach space*. A partially ordered real normed linear space which is also a Hilbert space is called a *partially ordered Hilbert space*.

A cone X_+ on a real normed linear space X is said to be generating, if $X = X_+ - X_+$, normal if there exists $\delta > 0$ such that $||x + y|| \geq \delta ||x||$ for all $x, y \in X_+$ and solid if $int(X_+) \neq \emptyset$, where $int(X_+)$ denotes the set of all interior points of X_+. If X is a Hilbert space, then a cone X_+ is said to be self-dual, if $X_+ = X_+^*$.

Let us include some examples. The real Euclidean space X with the usual norm and $X_+ = \mathbb{R}_+^n$, the nonnegative orthant is a solid, generating, and normal cone. The usual cones of nonnegative sequences in the space c (of all convergent real sequences), the space c_0 (of all convergent real sequences converging to zero) as subspaces of $\ell^p, 1 \leq p \leq \infty$, and the latter spaces are normal. Here the space ℓ^p is endowed with the usual cone

$$\ell_+^p := \{x = (x_1, x_2, \ldots) \in \ell^p : x_i \geq 0 \text{ for all } i\}.$$

ℓ_+^p is not solid for $1 \leq p < \infty$, while ℓ_+^∞ is a solid cone. Let $X = C(T)$ be the space of all real-valued continuous functions on a compact Hausdorff space T. Let X_+ be the cone of all nonnegative functions on $C(T)$. Then X_+ is solid, generating, and normal. The cone of all functions on $[0, 1]$ nonnegative almost everywhere in the space $L^p([0, 1])$ has properties similar to the case of ℓ^p. Consider the space $C^1([0, 2\pi])$ with the norm $\| f \| = \| f \|_\infty + \| f' \|_\infty, f \in C^1([0, 2\pi])$, where the prime denotes the derivative and both the norms on the right-hand side denote the supremum norm. Let X_+ be the cone of nonnegative functions which also belong to $C^1([0, 2\pi])$. Then X_+ is not a normal cone. Let us also list a few relationships between these specialized cones which will be useful in our discussion. If a cone in a normed linear space has nonempty interior, then it is generating. If a cone in a Banach space is generating, then its dual cone is normal in the dual space. For proofs of these results, we refer to [136].

Let us recall that if X is a partially ordered real normed linear space with the positive cone X_+, then an operator $T \in \mathcal{B}(X)$ is said to be X_+-nonnegative (nonnegative), if $TX_+ \subseteq X_+$. In this case, we write $T \geq 0$. Also, for $T_1, T_2 \in \mathcal{B}(X)$, by $T_1 \geq T_2$ we mean that $T_1 - T_2 \geq 0$. In order to view the results for the case of infinite dimensional spaces in a proper perspective, we consider a reformulation of Theorem 6.2.1. $X_+ = \mathbb{R}_+^n$ is a cone in X. The equivalence of statements (a) and (c) can now be paraphrased as: $s > \rho(B)$, with B nonnegative if and only if $sI - B$ is invertible and

$$(sI - B)^{-1}(X_+) \subseteq X_+$$

(or the same thing, $sI - B$ is inverse positive). Schaefer proved an analogue of Theorem 6.2.1 for operators over Banach spaces. Apparently, the term M-operators was not used. Let $\mathcal{B}(X)$ denote the space of all bounded linear operators on the normed linear space X. We then have the following (Proposition 2) [98]: Let X be a partially ordered Banach space with the positive cone X_+ such that X_+ and X_+^* are normal and $T \in \mathcal{B}(X)$ satisfy $T(X_+) \subseteq X_+$. Then $\alpha > \rho(T)$ if and only if

$$(\alpha I - T)^{-1} \in \mathcal{B}(X)$$

and

$$(\alpha I - T)^{-1}(X_+) \subseteq X_+.$$

The first use of the term M-operators in the context of infinite dimensional spaces seems to have been made by Marek and Szyld [71]. We recall this next. Let X be a partially ordered Banach space with the positive cone X_+. An operator $T \in \mathcal{B}(X)$ is said to be a Z-operator if $T = sI - P$, with $s \geq 0, P(X_+) \subseteq X_+$. A Z-operator is said to be an M-operator if $s \geq \rho(P)$. The set of all invertible M-operators will be denoted by $\mathbb{M}_{inv}$.

A bounded linear operator A on X is said to have property "p" if every number $\lambda \in \sigma(A)$ with $|\lambda| = \rho(A)$ is a pole of the resolvent operator $R(\mu, A) := (\mu I - A)^{-1}$.

Let us now recall a result of Marek and Szyld, which presents sufficient conditions under which an M-operator is inverse positive (Lemma 4.2) [71]. Let X be a partially ordered normed linear space with the positive cone X_+. Suppose that a bounded linear operator A on X could be written as $A = U - V$ with U, V also bounded and linear. Suppose, further that

$$U^{-1}V \geq 0 \text{ and } A^{-1}V \geq 0$$

and that both have property "p". Then $B = I - U^{-1}V$ is inverse positive.

There is at least one more definition of an M-operator (different from the one above and also a weaker version) that has been proposed in the infinite dimensional case by Kalauch (Chap. 2, Sect. 2.4, [55]). Both these reduce to the notion of an invertible M-matrix, when applied to the real Euclidean space and when the cone is taken to be the nonnegative orthant. For the reader interested in situations when these two concepts coincide, we refer to Corollary 2.4.5 in [55]. For our purposes, we consider the stronger version of the definition of an M-operator as given in [71]. For such operators, in Theorem 6.4.3, a generalization of Theorem 6.2.1 over Banach spaces is presented, and in Theorem 6.4.6, an extension of the result on intervals of M-matrices of [97] to operator intervals in Banach spaces.

As the third broad objective, the authors of [92] extended the definition of a Q-matrix to the infinite dimensional case. In fact, two classes of Q-operators were considered, one just named a Q-operator and another called a Q_s-operator. It is shown that for Z-operators, the former is stronger than the latter. More importantly, they demonstrated that a Z-operator which is also a Q-operator must be inverse positive. This is the main result of Sect. 6.5 and is proved in Theorem 6.5.1.

We organize this chapter as follows. In the next section, we consider P-operators and review certain important results for such operators, derived recently. In the fourth section, we prove some interesting results on M-operators. In the fifth section, we introduce the notion of Q-operators and study their relationship with M-operators. In the concluding section, we present a characterization theorem for the nonnegativity of the Moore–Penrose inverse of a Gram operator on a Hilbert space.

6.3 P-Operators

In this section, we make a review of some relationships between certain operators having the P-property and invertibility of some specific subsets of intervals of operators, that were proved recently. Certain partial results were reported in this work. The precise statements are given in this section. The matrix case has been investigated by Johnson and Tsatsomeros [50].

We begin with the notion of a P-operator (Definition 6.3.1, [92]). Let X be a real Banach space. A sequence $\{z^n : n \in \mathbb{N}\}$ in X is said to be a Schauder basis for X if for each $x \in X$, there exists a unique sequence of scalars $\{\alpha_n(x)\}$ such that $x = \sum_{n=1}^{\infty} \alpha_n(x) z^n$. In such a case, we denote $x_n = \alpha_n(x), n \in \mathbb{N}$. For the Banach spaces $l^p(\mathbb{N})$ with $1 \leq p < \infty$, by the standard Schauder basis we mean the set $\{e^n : n \in \mathbb{N}\}$, where e^i denotes the vector whose ith entry is one and all other entries are zero.

Definition 6.3.1. Let X be a Banach space with a Schauder basis. A bounded linear operator $T : X \longrightarrow X$ is said to be a P-operator relative to the given Schauder basis, if for any $x \in X$, the inequalities $x_i(Tx)_i \leq 0$ for all i imply that $x = 0$.

For $1 \leq p < \infty$, consider the linear operator $T : l^p(\mathbb{N}) \longrightarrow l^p(\mathbb{N})$ defined by

$$T(x) = (x_1, \frac{x_2}{2}, \frac{x_3}{3}, \dots), \; x \in l^p(\mathbb{N}).$$

Then T is a P-operator with respect to the standard Schauder basis. Let T be a positive definite operator on $l^2(\mathbb{N})$, viz., $\langle Tx, x \rangle > 0$ for all $0 \neq x \in l^2(\mathbb{N})$. Then T is a P-operator with respect to the standard Schauder basis. That the converse is not true is illustrated by the following example. Let $T : l^2(\mathbb{N}) \longrightarrow l^2(\mathbb{N})$ be defined by

$$T(x_1, x_2, x_3, \cdots) = (x_1 + x_2, x_2, x_3, \cdots), \; x \in l^2(\mathbb{N}).$$

Then T is a P-operator relative to the standard Schauder basis but not a positive definite operator.

In the finite dimensional case, the definition of a P-operator coincides with that of a P-matrix due to the sign non-reversal property. In this situation, A is a P-matrix if and only if A^{-1} exists and A^{-1} is also a P-matrix. But unlike the matrix case, over infinite dimensional spaces, a P-operator need not be invertible, for instance, the first example given above. However, we have the following result. Let $T \in \mathcal{B}(X)$

be an invertible P-operator relative to a given Schauder basis of X. Then T^{-1} is a P-operator relative to the same Schauder basis.

Next, we develop some notation. Let X be a Banach space with a Schauder basis $\mathbb{B}$. For $A, B \in \mathcal{B}(X)$, let

$$h(A,B) := \{C \in \mathcal{B}(X) : C = tA + (1-t)B,\ t \in [0,1]\}.$$

Let T be a fixed diagonal operator relative to $\mathbb{B}$ with diagonal entries in $[0, 1]$. Let

$$c(A,B) := \{C \in \mathcal{B}(X) : C = AT + B(I-T)\}$$

and

$$r(A,B) := \{C \in \mathcal{B}(X) : C = TA + (I-T)B\}.$$

Let $\mathcal{GL}(X)$ denote the set of bounded linear invertible operators on X. The next result gives a characterization for the inclusion $h(A,B) \subseteq \mathcal{GL}(X)$ to hold. This generalizes a matrix result (Observation 3.1, [50]). Let $A, B \in \mathcal{B}(X)$ be invertible. Then $h(A,B) \subseteq \mathcal{GL}(X)$ if and only if BA^{-1} has no negative spectral value.

For matrices, it is known that any matrix in the subset $r(A,B)$ is invertible if and only if BA^{-1} is a P-matrix [50]. In the next result, a partial generalization is considered. Let X be a Banach space with a Schauder basis $\mathbb{B} = \{z^n : n \in \mathbb{N}\}$ and $A, B \in \mathcal{B}(X)$ be invertible. If $r(A,B)$, relative to $\mathbb{B}$, satisfies $r(A,B) \subseteq \mathcal{GL}(X)$, then BA^{-1} is a P-operator with respect to $\mathbb{B}$. Conversely, if BA^{-1} is a P-operator with respect to $\mathbb{B}$, then $0 \notin \sigma_p(C)$ for all $C \in r(A,B)$.

In a similar way, the following result can be established.

Let X be a Banach space with a Schauder basis $\mathbb{B} = \{z^n : n \in \mathbb{N}\}$ and $A, B \in \mathcal{B}(X)$ be invertible. If $c(A,B)$, relative to $\mathbb{B}$, satisfies $c(A,B) \subseteq \mathcal{GL}(X)$, then $B^{-1}A$ is a P-operator relative to $\mathbb{B}$. Conversely, if $B^{-1}A$ is a P-operator relative to $\mathbb{B}$, then $0 \notin \sigma_p(C)$ for all $C \in c(A,B)$. For proofs, we refer to [92].

Let us conclude this section by pointing out to the fact that, just as in the matrix case, the Schur complement and the principal pivot transformation preserve the P-property of an operator. The proofs of these results are given in Sect. 3.2 of the work [92].

6.4 M-Operators

In this section, we consider M-operators. Our intention is to review an extension of Theorem 6.2.1 to infinite dimensional spaces, which was proved recently. Towards this objective, we introduce the notion of a Stiemke operator. Using this, we present the desired characterization in Theorem 6.4.3. Our next goal is to extend the interval result for M-matrices and this is given in Theorem 6.4.6, under certain conditions on the cone. We would like to remark that in a recent work of Sivakumar and Weber

(Theorem 6.4.2, [126]), necessary and sufficient conditions were presented for an operator interval to contain only invertible operators. This interval is more general (and much larger) than what we have considered in Theorem 6.4.6. However, the result of [126] is applicable for rather restricted classes of cones. Let us also add that whereas their proof uses the notion of splittings of operators, the argument in the proof of Theorem 6.4.6 uses only the monotonicity of the spectral radius.

We start with the definition of a Steimke operator.

Definition 6.4.1 (Definition 4.1, [92]). Let X be a partially normed linear space with a solid cone X_+. An operator $A \in \mathcal{B}(H)$ is called a Stiemke operator if $A(int(X_+)) \cap int(X_+) \neq \emptyset$.

In the next result, we provide a sufficient condition for an operator to be a Stiemke operator.

Theorem 6.4.2 (Theorem 4.1, [92]). *Let X be a partially ordered normed linear space where the positive cone X_+ is solid. If $A \in \mathcal{B}(X)$ is invertible and $A^{-1} \geq 0$, then A is a Stiemke operator.*

Proof. Let $q \in int(X_+)$ and $p = A^{-1}q$. Then $p \in X_+$ and $Ap \in int(X_+)$. Now, we shall show that $p \in int(X_+)$. Since $Ap \in int(X_+)$, there exists $\epsilon > 0$ such that the open ball (with center Ap and radius ϵ) $\mathcal{U}(Ap, \epsilon) \subseteq X_+$. Since A is bounded there exists $\delta > 0$ such that $A(\mathcal{U}(p, \delta)) \subseteq \mathcal{U}(Ap, \epsilon)$. Since $A^{-1}(X_+) \subseteq X_+$, it follows that $\mathcal{U}(p, \delta) \subseteq A^{-1}(\mathcal{U}(Ap, \epsilon)) \subseteq A^{-1}(X_+) \subseteq X_+$. Thus $p \in int(X_+)$. Hence $A(int(X_+)) \cap int(X_+) \neq \emptyset$ and so A is a Stiemke operator. □

We are now in a position to discuss the main result of this section. This result presents two equivalent conditions for a Z-operator to be an invertible M-operator. Part of this result was already established by Schaefer [98], as recorded earlier. As mentioned there, this extends Theorem 6.2.1 to infinite dimensional spaces.

Theorem 6.4.3 (Theorem 4.2, [92]). *Let X be a Banach space, X_+ be a solid and normal cone, and A be a Z-operator. Then the following statements are equivalent:*

(a) $A \in \mathbb{M}_{inv}$.
(b) *A is an invertible Stiemke operator.*
(c) *A is invertible and $A^{-1} \geq 0$.*

Let us consider a few examples, next.

Example 6.4.4. Consider the Hilbert space $H = \mathbb{R} \oplus l^2(\mathbb{N}) = \{(\xi, x) : \xi \in \mathbb{R}, x \in l^2(\mathbb{N})\}$ with the inner product defined in the following way:

$$\langle(\xi_1, x_1), (\xi_2, x_2)\rangle = \xi_1\xi_2 + \langle x_1, x_2\rangle.$$

Then the norm induced by the inner product is $||(\xi, x)|| = \sqrt{\xi^2 + ||x||^2}$. Consider the cone

$$H_+ = \{(\xi, x) \in H : \xi \geq 0, \xi \geq ||x||\}$$

on H called the hyperbolic cone [69]. This is an extension of the definition of the ice-cream cone to infinite dimensional normed linear spaces. It may be shown that $int(H_+) \neq \emptyset$. Let $B : l^2(\mathbb{N}) \to l^2(\mathbb{N})$ be defined by

$$B(x) = (t_1x_1, t_2x_2, t_3x_3, \cdots) \text{ with } \frac{1}{2} < t_i \leq 1 \text{ for all } i.$$

Consider the operator D on H defined by $D = \begin{pmatrix} 1 & 0 \\ 0 & B \end{pmatrix}$. Then D is a bounded linear operator on H. Also, $||Bx|| \leq ||x||$ for all $x \in l^2(\mathbb{N})$. If $(\xi, x) \in H_+$, then $\xi \geq ||x||$ so that $\xi \geq ||Bx||$. Hence D is nonnegative, i.e., $D(H_+) \subseteq H_+$. It is clear that $\rho(D) \leq 1$. Define $A : H \longrightarrow H$ by $A = 2I - D$. Then $A \in \mathbb{M}_{inv}$.

Example 6.4.5. Consider the Banach Space $X = l^\infty(\mathbb{N})$ and the cone

$$X_+ = \{x = (x_i) \in X : x_i \geq 0 \text{ for all } i\}.$$

Then X_+ is solid and normal. Consider the operator T whose entries (t_{ij}), relative to the standard Schauder basis $\{e^1, e^2, \cdots\}$, satisfy the conditions $t_{ij} \geq 0$ and $\sum_{j=1}^{\infty} t_{ij} \leq 1$. Then $T \in \mathcal{B}(X)$ and $T(X_+) \subseteq X_+$. It is clear that $e = (1, 1, 1, \cdots) \in int(X_+)$ and $Te \leq e$. We then have $\rho(T) \leq 1$. Thus $\lambda I - T \in M_{inv}$ for all $\lambda > 1$.

Next, we turn our attention to intervals of operators. We begin with the following definition. Let X be a partially ordered Banach space and X_+ be the positive cone in X. For $A, B \in \mathcal{B}(X)$, with $A \leq B$, define $J(A, B)$, an interval of operators by

$$J(A, B) = [A, B] = \{D \in \mathcal{B}(X) : A \leq D \leq B\}.$$

The following result is an extension of the result of Rohn for an interval of M-matrices:

Theorem 6.4.6 (Theorem 4.3, [92]). *Let X be a Banach space and X_+ be a normal and generating cone in X. Let $A, B \in \mathcal{B}(X)$ with $A \leq B$. Then $A, B \in \mathbb{M}_{inv}$ if and only if $J(A, B) \subseteq \mathbb{M}_{inv}$.*

Proof. Sufficiency is obvious. We prove the necessity part. Let $D \in J$. Consider $A = sI - E, B = sI - F$ with $E \geq 0, F \geq 0, s > \rho(E)$ and $s > \rho(F)$. It follows that D can be represented as $D = sI - G$ with $G \geq 0$. Since $E \geq G \geq F \geq 0$, by the monotonicity of the spectral radius, we have $\rho(E) \geq \rho(G) \geq \rho(F)$, so that $s > \rho(G)$. Hence $D \in \mathbb{M}_{inv}$. □

Remark 6.4.7. The result of [126] mentioned in the beginning of this section is applicable for cones with nonempty interior. This assumption is not made in Theorem 6.4.6. The assumptions that X_+ is generating and normal are indispensable in Theorem 6.4.6, as shown in [92].

6.5 Q-Operators

In this section, we recall the concept of Q-operators as introduced in [92] and their relationships with M-operators. Let us recall that in the matrix case, the statement that a Z-matrix is a Stiemke matrix is equivalent to the condition that the given matrix is inverse positive or that it is a Q-matrix. This result is quite well known in the theory of linear complementarity problems [11, 21]. We show that a Q-operator which is also a Z-operator must be an (invertible) M-operator, so that it is inverse positive. This statement is included in Theorem 6.5.1, which is the main result of this section.

Let H be a real Hilbert space, H_+ be a cone in H, and H_+^* denote the dual cone of H_+. For a given vector $q \in H$ and an operator $A \in \mathcal{B}(H)$ the cone complementarity problem, written as $LCP(A, q)$, is to find a vector $z \in H_+$ such that $Az+q \in H_+^*$ and $\langle z, Az + q\rangle = 0$. By taking $z = 0$, it follows that $LCP(A, q)$ has solution whenever $q \in H_+^*$. For $H = \mathbb{R}^n$, $H_+ = \mathbb{R}^n_+$ and $q \in \mathbb{R}^n$ the cone complementarity problem reduces to the linear complementarity problem.

The definition of a Q-operator is now immediate, as in the case of matrices. An operator $A \in \mathcal{B}(H)$ is said to be a Q-operator if $LCP(A, q)$ has a solution for all $q \in H$.

In the finite dimensional case with the usual cone, the definition of Q-operators coincides with the notion of Q-matrices. Consider the Hilbert space $l^2(\mathbb{N})$ with the usual cone $l^2(\mathbb{N})_+$. Then $l^2(\mathbb{N})_+^* = l^2(\mathbb{N})_+$. The identity operator I on $l^2(\mathbb{N})$ is a Q-operator. More generally, any "diagonal" operator on $l^2(\mathbb{N})$ whose diagonals are all greater than or equal to 1 can be shown to be a Q-operator.

Next, we consider a subclass of Stiemke operators (Definition 6.4.1), in this instance, over Hilbert spaces. Let H be a real Hilbert space with a solid cone H_+ and A be a Stiemke operator. Then A is called a Q_s-*operator* if for some $q \in int(H_+)$ with $Aq \in int(H_+)$ the problems $LCP(A^*, tq)$ for $t = 0, 1$ have only one solution, viz., zero.

Let us consider an example. Let $H = \mathbb{R} \oplus l^2(\mathbb{N}) = \{(\xi, x) : \xi \in \mathbb{R}, x \in l^2(\mathbb{N})\}$ and H_+ be as in Example 6.4.4. Let D be defined on H by $D = \begin{pmatrix} 1 & 0 \\ 0 & B \end{pmatrix}$, where $B : l^2(\mathbb{N}) \to l^2(\mathbb{N})$ is given by $B = diag(\frac{1}{2}, \frac{1}{3}, \frac{1}{4}, \cdots)$. Then D is a Q_s-operator. We omit the details of the proof.

For any Z-matrix A, it is not difficult to show that if A is a Q_s-matrix, then it is a Q-matrix. Now, we turn to the main result of this section as well as this chapter. As mentioned in the introduction, this is motivated by Theorem 7 in [40]. As this is an important result, we include a complete proof.

Theorem 6.5.1 (Theorem 5.4, [92]). *Let H be a real Hilbert space with a solid, normal, and self-dual cone H_+. Let $A \in \mathcal{B}(H)$ be a Z-operator. Consider the following statements:*

(*a*) $A \in \mathbb{M}_{inv}$.
(*b*) *A is an invertible Stiemke operator.*

(c) A is invertible and $A^{-1} \geq 0$.
(d) A is injective and is a Q-operator.
(e) A is injective and is a Q_s*-operator.*

Then we have (a) $\Leftrightarrow$ *(b)* $\Leftrightarrow$ *(c) and (d)* $\Rightarrow$ *(c)* $\Rightarrow$ *(e).*

Proof. The equivalence of (*a*), (*b*), and (*c*) follows from Theorem 6.4.3.

(*d*) $\Longrightarrow$ (*c*) : Let $q \in H_+$ and consider $LCP(A, -q)$. Since A has the Q-property, there exists $x \in H_+$ such that

$$x \in H_+,\ y = Ax - q \in H_+ \text{ and } \langle x, y \rangle = 0.$$

Since A is a Z-operator, we have $A = sI - T$ where $s \geq 0$ and $T \geq 0$. Consider

$$\langle Ax, y \rangle = \langle (sI - T)x, y \rangle = s\langle x, y \rangle - \langle Tx, y \rangle.$$

Since $TH_+ \subseteq H_+$, H_+ is self-dual and $\langle x, y \rangle = 0$, we have $\langle Ax, y \rangle \leq 0$. So, $0 \geq \langle Ax, y \rangle = \langle y + q, y \rangle = ||y||^2 + \langle q, y \rangle$. The second term is nonnegative since H_+ is self-dual. This means that $y = 0$ and so $Ax = q$. What we have shown is that for every $q \in H_+$ there exists $x \in H_+$ such that $q = Ax$. Since H_+ is generating, this implies that A is surjective. Already, A is injective. Hence, A^{-1} exists and so $A \in \mathbb{M}_{inv}$. It now follows that $A^{-1} \geq 0$.

(*c*) $\Longrightarrow$ (*e*) : Since (c) is equivalent to (b), A is an invertible Stiemke operator. Now, let $q \in int(H_+)$ with $Aq \in int(H_+)$. Let u_t be a solution for $LCP(A^*, tq)$ for $t = 0, 1$. Then

$$u_t \in H_+, v_t := A^* u_t + tq \in H_+ \text{ and } \langle u_t, v_t \rangle = 0.$$

Since A is a Z-operator, as shown earlier, we have $\langle Av_t, u_t \rangle \leq 0$. Thus, $0 \geq \langle Av_t, u_t \rangle = \langle v_t, A^* u_t \rangle = \langle A^* u_t + tq, A^* u_t \rangle = \| A^* u_t \|^2 + t\langle Aq, u_t \rangle$. Since both the terms on the extreme right-hand side are nonnegative, we conclude that $A^* u_t = 0$. Since A is invertible, we then have $u_t = 0$. Thus A is a Q_s-operator.

□

The concluding result of this chapter is an important consequence of Theorem 6.5.1, where we obtain the following well-known result for Z-matrices. We omit the details and refer the reader to [92].

Corollary 6.5.2. *Let* $A \in \mathbb{R}^{n \times n}$ *be a Z-matrix. Then the following statements are equivalent:*

(a) $A \in \mathbb{M}_{inv}$.
(b) There exists $p > 0$ *such that* $Ap > 0$.
(c) A is invertible and $A^{-1} \geq 0$.
(d) A is a Q-matrix.

6.6 Nonnegative Moore–Penrose Inverses of Gram Operators

Let us recall that a matrix $A \in \mathbb{R}^{n\times n}$ is called *monotone* if

$$Ax \geq 0 \Longrightarrow x \geq 0,$$

where the ordering of vectors is the usual component wise order. Collatz proposed this notion and studied monotone matrices in his investigations on solving certain differential equations by applying finite difference methods. He studied many properties of such matrices and gave sufficient conditions for a matrix to be monotone. In particular, he showed that A is monotone if and only if A is invertible and that the inverse of A is (entrywise) nonnegative. In other words, a monotone matrix is inverse positive in the terminology that we had adopted in Sect. 2.1, Chap. 2. As discussed there, any M-matrix is monotone. One of the class of monotone matrices that has received considerable attention in recent years is the class of Gram matrices of a given matrix. Recall that for $A \in \mathbb{R}^{m\times n}$, the matrix A^*A (or AA^*) is referred to as the *Gram* matrix corresponding to A. These matrices are important in convex optimization problems and a lot of attention has been paid to their characterizations. In this section, our intention will be to review a rather recent work of Kurmayya and Sivakumar [61], where some of the characterizations for monotonicity of Gram matrices have been extended to the case of Gram operators between Hilbert spaces. Inverse positivity is extended to the nonnegativity of the Moore–Penrose inverse.

In order to state their result, we need some preliminaries. A cone C in an inner product space is said to be *acute* if $\langle x, y\rangle \geq 0$ for all $x, y \in C$. C is said to be *obtuse*, if $C \cap \overline{sp\,\{C\}}$ is acute, where $sp\,\{C\}$ denotes the linear subspace spanned by C and $\overline{sp\,\{C\}}$ denotes is topological closure. The lexicographic cone in ℓ^2 is an example of a (nonclosed) acute cone. In particular, if H and K are real Hilbert spaces, $A \in \mathcal{B}(H, K)$ and $C = AH_+$, where H_+ is a cone in H, then the obtuseness of C is the same as the acuteness of $C^* \cap R(A)$, where as before, C^* denotes the dual cone of C. For $A \in \mathcal{B}(H, K)$ we shall refer to A^*A as the Gram operator corresponding to A.

We are now in a position to state the aforementioned characterization. Let $A \in \mathcal{B}(H, K)$ be such that $R(A)$ is closed. Let $H_+ \subseteq H$ be a closed cone such that $A^\dagger A(H_+) \subseteq H_+$. Set $C = AH_+$ and $D = (A^\dagger)^* H_+^*$. Then Kurmayya and Sivakumar (Theorem 6.4.6, [61]) show that the following statements are equivalent:

(*a*) $(A^*A)^\dagger(H_+^*) \subseteq H_+$.
(*b*) $C^* \cap R(A) \subseteq C$.
(*c*) D is acute.
(*d*) C is obtuse.
(*e*) $A^*Ax \in P_{R(A^*)}(H_+^*)$, $x \in R(A^*) \Longrightarrow x \in H_+$.
(*f*) $A^*Ax \in H_+^*$, $x \in R(A^*) \Longrightarrow x \in H_+$.

The following consequences are obtained. For $H = \mathbb{R}^n$, $K = \mathbb{R}^m$ and $H_+ = \mathbb{R}^n_+ \cap R(A)$, the conditions (*b*)–(*f*) above are equivalent to the statement (Corollary 3.7, [61]):

$$(A^*A)^{\dagger}(\mathbb{R}^n_+) \subseteq \mathbb{R}^n_+,$$

viz., the (entrywise) nonnegativity of $(A^*A)^{\dagger}$. It must be mentioned that, in addition to the assumptions made as above, if the matrix A has a full column rank, then the conditions (*b*)–(*f*) above are equivalent to the statement (Corollary 3.8, [61]):

$$(A^*A)^{-1} \geq 0.$$

It may be verified that some prominently known characterizations of inverse positivity follow as a particular case of this result. Sivakumar also obtained a new characterization of the nonnegativity of the Moore–Penrose inverse of Gram operators (Theorem 13.7, [125]) using a completely different proof approach. Let us conclude this section by briefly mentioning that, in that work, it is shown that an additional statement of equivalence for the inclusion $(A^*A)^{\dagger}(H^*_+) \subseteq H_+$ to hold is given by:

for every $x \in H$, there exists $y \in K$ such that $y \pm x \in AH_+$ with $\| y \| \leq \| x \|$.

Chapter 7
Infinite Linear Programming

7.1 Introduction

Infinite linear programming problems are linear optimization problems where, in general, there are infinitely (possibly uncountably) many variables and constraints related linearly. There are many problems arising from real world situations that can be modelled as infinite linear programs. These include the bottleneck problem of Bellman in economics, infinite games, and continuous network flows, to name a few. We refer to the excellent book by Anderson and Nash [2] for an exposition of infinite linear programs, a simplex type method of solution and applications. Semi-infinite linear programs are a subclass of infinite programs, wherein the number of variables is finite with infinitely many constraints in the form of equations or inequalities. Semi-infinite programs have been shown to have applications in a number of areas that include robot trajectory planning, eigenvalue computations, vibrating membrane problems, and statistical design problems. For more details we refer to the two excellent reviews by Hettich and Kortanek [43] and Polak [90] on semi-infinite programming.

We shall be making use of the notion of a partially ordered Hilbert space, which was discussed in Chap. 6. Let H_1 and H_2 be real Hilbert spaces with H_2 partially ordered, $A : H_1 \to H_2$ be bounded linear, $c \in H_1$ and $a, b \in H_2$ with $a \leq b$. Consider the linear programming problem called *interval linear program* $ILP(a, b, c, A)$:

$$\text{Maximize } \langle c, x \rangle$$
$$\text{subject to } a \leq Ax \leq b.$$

Such problems were investigated by Ben-Israel and Charnes [9] in the case of finite dimensional real Euclidean spaces and explicit optimal solutions were obtained using the techniques of generalized inverses of matrices. These results were later extended to the general case and several algorithms were proposed for interval

P.N. Shivakumar et al., *Infinite Matrices and Their Recent Applications*,
DOI 10.1007/978-3-319-30180-8_7

linear programs. Some of these results were extended to certain infinite dimensional spaces by Kulkarni and Sivakumar [59, 60].

The objective of this chapter is to present an algorithm for the countable semi-infinite interval linear program denoted $SILP(a, b, c, A)$ of the type:

$$\text{Maximize } \langle c, x\rangle$$
$$\text{subject to a} \leq Ax \leq b,$$

where $c \in \mathbb{R}^m$, H is a real separable partially ordered Hilbert space, $a, b \in H$ with $a \leq b$ and $A : \mathbb{R}^m \longrightarrow H$ is a (bounded) linear map. Finite dimensional truncations of the above problems are solved as finite dimensional interval programs. This generates a sequence $\{x^k\}$ with each x^k being optimal for the truncated problem at the kth stage. This sequence is shown to converge to an optimal solution of the problem $SILP(a, b, c, A)$. We also show how this idea can be used to obtain optimal solutions for continuous semi-infinite linear programs and to obtain approximate optimal solutions to doubly infinite interval linear programs.

7.2 Preliminaries

Let us briefly recall the necessary definitions and results that are needed in the subsequent discussion. We will state our results generally and then specialize later. Let H_1 and H_2 be real Hilbert spaces with H_2 partially ordered, $A : H_1 \rightarrow H_2$ be linear, $c \in H_1$ and $a, b \in H_2$ with $a \leq b$. Consider the interval linear program $ILP(a, b, c, A)$:

$$\text{Maximize } \langle c, x\rangle$$
$$\text{subject to } a \leq Ax \leq b.$$

Definition 7.2.1. Set $F = \{x : a \leq Ax \leq b\}$. A vector x^* is said to be feasible for $ILP(a, b, c, A)$ if $x^* \in F$. The problem $ILP(a, b, c, A)$ is said to be feasible if $F \neq \emptyset$. A feasible vector x^* is said to be optimal if $\langle c, (x^* - x)\rangle \geq 0$ for every $x \in F$. $ILP(a, b, c, A)$ is said to be bounded if $sup\{\langle c, x\rangle : x \in F\} < \infty$. If $ILP(a, b, c, A)$ is bounded, then the value α of the problem is defined as $\alpha = sup\{\langle c, x\rangle : x \in F\}$.

We will assume throughout the paper that $F \neq \emptyset$. Boundedness is then characterized by Lemma 7.2.2, to follow. Let $A : H_1 \longrightarrow H_2$ be linear. A linear map $T : H_2 \longrightarrow H_1$ is called a $\{1\}$-inverse of A if $ATA = A$. One may recall that this equation is just one of the equations that the Moore–Penrose inverse is required to satisfy and so the Moore–Penrose inverse of A is one choice for a $\{1\}$-inverse. In this regard, let us also remember that a bounded linear map A has a bounded $\{1\}$-inverse if $R(A)$, the range space of A is closed [77].

Lemma 7.2.2 (Lemma 2, [59]). *Suppose H_1 and H_2 are real Hilbert spaces, H_2 is partially ordered, and ILP(a,b,c,A) is feasible. If ILP(a,b,c,A) is bounded then $c \perp N(A)$. The converse holds if A has a bounded $\{1\}$-inverse and $[a,b] = \{z \in H_2 : a \leq z \leq b\}$ is bounded.*

7.3 Finite Dimensional Approximation Scheme

Let H be a partially ordered real Hilbert space with an orthonormal basis $\{u^n : n \in \mathbb{N}\}$ such that $u^n \in H_+$ for all n, where H_+ is the positive cone in H. Let $\{S_n\}$ be a sequence of subspaces in H such that $\{u^1, u^2, \cdots, u^n\}$ is an orthonormal basis for S_n. Then $\overline{\cup_{n=1}^{\infty} S_n} = H$. Let $T_n : H \longrightarrow H$ be the orthogonal projection of H onto S_n. Then for $x = (x_1, x_2, \cdots) \in H$, one has

$$T_n(x) = (x_1, x_2, \cdots, x_n, 0, 0, \cdots).$$

Then the sequence $\{T_n\}$ converges strongly to the identity operator on H.

Consider the semi-infinite linear program *SILP*(a,b,c,A). Suppose that this problem is bounded and that the columns of A are linearly independent. Then *SILP*(a,b,c,A) is bounded, by Lemma 7.2.2. Define the kth subproblem *SILP*$(T_k a, T_k b, c, T_k A)$ denoted by *SILP*$_k$ by:

$$\begin{aligned} &\text{Maximize} && \langle c, x\rangle \\ &\text{subject to} && T_k a \leq T_k A x \leq T_k b. \end{aligned}$$

SILP$_k$ has k interval inequalities in the m unknowns $x_i, i = 1, 2, \cdots, m$. In view of the remarks given above, it follows that *SILP*$(T_k a, T_k b, c, T_k A)$ is bounded whenever $k \geq m$.

We now state a convergence result for the sequence of optimal solutions of the finite dimensional subproblems. The proof will appear elsewhere [53].

Theorem 7.3.1. *Let $\{x^k\}$ be a sequence of optimal solutions for SILP$_k$. Then $\{x^k\}$ converges to an optimal solution of SILP(a,b,c,A).*

Next, we present a simple example to illustrate Theorem 7.3.1. A detailed study of various examples showing how the above theorem can be used to obtain optimal solutions even for a continuous semi-infinite linear program is presently underway [53].

Example. Let $H = \ell^2, A : \mathbb{R}^2 \to \ell^2$ be defined by

$$A = \begin{pmatrix} 1 & 0 \\ 1 & 1 \\ \frac{1}{2} & \frac{1}{4} \\ \frac{1}{3} & \frac{1}{9} \\ \vdots & \vdots \end{pmatrix}$$

$a = (0, -1, -\frac{1}{4}, -\frac{1}{9}, \cdots), b = (1, 0, 0, \cdots)$, and $c = (1, 0)$. Consider the semi-infinite program $SILP(a, b, c, A)$ given by

$$\text{Maximize } x_1$$

$$\text{subject to } 0 \leq x_1 \leq 1,$$

$$\frac{-1}{k^2} \leq \frac{x_1}{k} + \frac{x_2}{k^2} \leq 0, \quad k = 1, 2, 3, \cdots \tag{7.1}$$

Clearly, $c \perp N(A)$, as $N(A) = \{0\}$ and so the problem is bounded by Lemma 7.2.2. Rewriting the second set of inequalities, we get

$$\frac{-1}{k} \leq x_1 + \frac{x_2}{k} \leq 0, k = 1, 2, 3, \cdots$$

Then $x_1 = 0$, by passing to limit as $k \to \infty$. Thus $(0, \alpha), -1 \leq \alpha \leq 0$ is an optimal solution for $SILP(a, b, c, A)$ with optimal value 0.

Now we consider the finite truncations. Let $S_k = \{e^1, e^2, \cdots, e^k\}$ where $\{e^1, e^2, e^3, \cdots\}$ is the standard orthonormal basis for ℓ^2. Define

$$T_k = \begin{pmatrix} I_k & 0 \\ 0 & 0 \end{pmatrix}$$

where I_k is the identity matrix of order $k \times k$, $k \geq 2$. Thus $R(T_k A) = S_k$. The subproblem $SILP(T_k a, T_k b, c, T_k A)$ is

$$\textit{Maximize } x_1$$

$$\text{subject to } 0 \leq x_1 \leq 1,$$

$$\frac{-1}{(l-1)^2} \leq \frac{x_1}{l-1} + \frac{x_2}{(l-1)^2} \leq 0, \; l = 2, 3, \cdots, k.$$

Clearly, the subproblem $SILP(T_k a, T_k b, c, T_k A)$ is bounded and the optimal solution is $x^k = (\frac{1}{k-2}, \frac{-(k-1)}{k-2})$. This converges to $(0, -1)$ in the limit as $k \to \infty$.

7.4 Approximate Optimal Solutions to a Doubly Infinite Linear Program

In this section, we show how we can obtain approximate optimal solutions to continuous infinite linear programming problems. With the notation adopted as earlier, assume that $ILP(a, b, c, A)$ is bounded. Let $c^{(k)}$ denote the vector in H whose first k components are the same as the vector c and whose other components are zero. Let A_k denote the matrix with k columns and infinitely many rows whose k columns are precisely the same as the first k columns of A in that order. Consider the problem:

$$\begin{array}{ll} \text{Maximize} & < c^{(k)}, u > \\ \text{subject to} & a \leq A_k u \leq b, \end{array}$$

for $u \in \mathbb{R}^k$. This problem is $SILP(a, b, c^{(k)}, A_k)$ for each k. Suppose that the columns of A_k are linearly independent for all k. Denoting $SILP(a, b, c^{(k)}, A_k)$ by P_k, we solve P_k using Theorem 7.3.1. Let $u^k = (u_1, u_2, \cdots, u_k)^T$ be an optimal solution for P_k. Let $u = (u^k, 0)^T \in H$. Then $A_{k+1}u = A_k u^k$. Since $a \leq A_{k+1}u \leq b$, the vector $u \in H$ is feasible for P_{k+1}. Let $v(P_k)$ denote the value of problem P_k, viz.,

$$v(P_k) := sup\{\langle c^{(k)}, u\rangle : a \leq A_k u \leq b.\}$$

Then

$$v(P_{k+1}) \geq v(P_k) \text{ as } \langle c^{(k+1)}, u\rangle \geq \langle c^{(k)}, u\rangle.$$

Let $x^k = (u^k, 0, 0 \cdots) \in H$. Then x^k is feasible for $ILP(a, b, c, A)$ and $\langle c, x^k\rangle = \langle c^{(k)}, x^k\rangle$, the value of P_k. Thus the sequence $\{\langle c, x^k\rangle\}$ is an increasing sequence bounded above by the value of $ILP(a, b, c, A)$ and is hence convergent. It follows that $\{x^k\}$ is weakly convergent. However, unlike the semi-infinite case, it need not be convergent. (In the following example, it turns out that $\{x^k\}$ is convergent.) So, we have optimal value convergence but not optimal solution convergence. Hence our method yields only an approximate optimal solution for a continuous linear program. It will be interesting to study how good is this approximation.

Example. Let $A : \ell^2 \to \ell^2$ be defined by

$$A = \begin{pmatrix} 1 & 0 & 0 & 0 \\ 1 & \frac{1}{2} & 0 & \cdots \\ 1 & \frac{1}{2} & \frac{1}{3} & \cdots \\ \vdots & \vdots & \vdots & \ddots \end{pmatrix}.$$

Let $c = (1, \frac{1}{4}, \frac{1}{9}, \cdots)$, $a = \mathbf{0}$, and $b = (1, \frac{1}{2}, \frac{1}{3}, \cdots)$. The problem $ILP(a, b, c, A)$ is

$$\text{Maximize } x_1 + \frac{x_2}{4} + \frac{x_3}{9} + \cdots$$

$$\text{subject to } 0 \leq \sum_{i=1}^{l} \frac{x_i}{i} \leq \frac{1}{l},\ l = 1, 2, 3, \cdots.$$

Clearly, $A \in BL(\ell^2)$ and $N(A) = \{0\}$. Therefore $c \perp N(A)$, i.e., the problem $ILP(a, b, c, A)$ is bounded. Consider the kth subproblem P_k:

$$\text{Maximize} \qquad \langle c^{(k)}, u \rangle$$
$$\text{subject to} \qquad a \leq A_r u \leq b, \ \ r = 1, 2, \cdots, k$$

where $c^{(k)} = (1, 1/4, 1/9, \cdots, 1/k^2, 0, 0, \ldots)$, and $A_k : \mathbb{R}^k \to \ell^2$ is defined by

$$A = \begin{pmatrix} 1 & 0 & \cdots & 0 \\ 1 & \frac{1}{2} & \cdots & \cdots \\ \vdots & \vdots & \cdots & \vdots \\ 1 & \frac{1}{2} & \cdots & \frac{1}{k} \\ \vdots & \vdots & \cdots & \vdots \end{pmatrix}.$$

Clearly, the subproblem is bounded. By the finite dimensional scheme, the optimal solution of the subproblem P_k is found to be

$$x^{(k)} = (1, -1, -\frac{1}{2}, -\frac{1}{3}, \cdots, -\frac{1}{k-1}, 0, 0, \ldots)$$

which converges to $x^* = (1, -1, -\frac{1}{2}, -\frac{1}{3}, \cdots)$. The optimal value converges to 6.450. We conclude by observing that since A is invertible, it is possible to solve the original problem by the methods in [59], directly.

Chapter 8
Applications

8.1 Introduction

In this chapter, we collect many applications of the various ideas that were discussed in the earlier chapters. In the second section, we show how the results for weakly chained diagonally dominant matrices, discussed in Sect. 2.3, have been applied in obtaining bounds for the ℓ^1 norm for the solutions of certain differential systems and also in deriving bounds for a critical parameter in electric circuit design. In Sect. 8.3, we review how a mapping problem could be reduced to an infinite system of linear equations and then solved. In Sect. 8.4, a similar idea is employed to show how the problem of the flow of fluids in and between two pipes could be handled. In the next section, viz., Sect. 8.5, we recall how some special double points of the Mathieu differential equation could be computed using techniques from infinite matrices. In Sect. 8.6, we discuss how the iterative method described earlier could be applied to obtain good approximate values of Bessel functions in certain intervals. Section 8.7 reviews results for the minimal eigenvalue of the Dirichlet Laplacian in an annulus. In the next section, an approximate solution providing the best match for the hydraulic head in a porous medium is presented. The next section, namely Sect. 8.9, considers eigenvalues of the Laplacian in an elliptic domain. The penultimate section studies the problem of the possibility of "hearing" the shape of a drum. The concluding section, Sect. 8.11, discusses how one could determine the zeros of a Taylor series.

8.2 Two Applications of Weakly Chained Diagonally Dominant Matrices

Let us recall that, in Sect. 2.3 the notion of weakly chained diagonal dominance for certain classes of Z-matrices was discussed. Results for upper and lower bounds for the minimal eigenvalue of A, and its corresponding eigenvector, and for the entries

P.N. Shivakumar et al., *Infinite Matrices and Their Recent Applications*,
DOI 10.1007/978-3-319-30180-8_8

of the inverse of A were reviewed there. In what follows, we show how these results have been applied to find meaningful two-sided bounds for both the ℓ^1-norm and the weighted Perron-norms of the solution $\mathbf{x}(t)$ to the linear differential system

$$\dot{\mathbf{x}} = -A\mathbf{x},\ \mathbf{x}(0) = \mathbf{x}^0 > 0.$$

Note that $\mathbf{x}(t) = (x_1(t), x_2(t), \ldots, x_n(t))^T$ and $\dot{\mathbf{x}} = \dot{\mathbf{x}}(t) = (\frac{dx_1(t)}{dt}, \ldots, \frac{dx_n(t)}{dt})^T$. In particular, these systems occur in R-C electrical circuits and a detailed analysis of a model for the transient behavior of digital circuits is studied by Shivakumar, Williams, Ye and Marinov [120].

In order to apply the results of Sect. 2.3, we assume that A is an irreducible weakly chained diagonally dominant Z-matrix with positive diagonal entries. It is proved (Theorem 5.1, [120]) that for all $t \geq 0$,

$$\sum_{i=1}^{n} z_i \mathbf{x}_i(t) = e^{-qt} \sum_{i=1}^{n} z_i \mathbf{x}_i^0, \tag{8.1}$$

where $q = q(A)$ is the Perron root and $z = (z_1, z_2, \ldots, z_n)^T$ is the unique normalized positive eigenvector (the Perron vector) of A^T corresponding to q.

If bounds for the Perron root are available, viz., $0 < q_m \leq q \leq q_M$, then one has

$$\frac{z_{\min}}{z_{\max}} \|\mathbf{x}_0\| e^{-q_M} \leq \|\mathbf{x}(t)\| \leq \frac{z_{\max}}{z_{\min}} \|\mathbf{x}_0\| e^{-q_m}, \tag{8.2}$$

where $z_{min} \leq z_i \leq z_{max}$ for all $i = 1, 2, \ldots, n$ and $\| \ . \ \|$ denotes the 1-norm. References to results of similar nature which motivated the above are given in [120].

Let us turn our attention to a problem in electrical circuits which was also considered by the same authors. It is rather well-known that if $\mathbf{v}(t) = (v_1(t), v_2(t), \ldots, v_n(t))^T$ denotes the vector of node voltages, then under certain conditions, the transient evolution of an R-C circuit is governed by the differential equation

$$C\frac{d\mathbf{v}(t)}{dt} = -G\mathbf{v} + g, \tag{8.3}$$

where C is a diagonal matrix with nonzero diagonal entries, g is a given vector, and G is a given matrix of conductances. If $\mathbf{v}^\infty$ is the so-called stationary regime voltage vector and if we set $\mathbf{x}(t) = \mathbf{v}(t) - \mathbf{v}^\infty$, then the differential equation given above reduces to

$$\frac{d\mathbf{x}(t)}{dt} = -C^{-1}G\mathbf{x}(t). \tag{8.4}$$

The matrix $C^{-1}G$ turns out to be an irreducible Z-matrix with positive diagonal entries with certain further properties. The crucial performance of the digital circuit is the high operating speed that is measured by the quantity

$$T(\epsilon) = \sup \left\{ t : \frac{\|\mathbf{x}(t)\|}{\|\mathbf{x}_0\|} = \epsilon \right\}, \tag{8.5}$$

where again, the norm is the 1-norm and $0 < \epsilon < 1$. In practice one takes the value $\epsilon = 0.1$. Finding an appropriate value of $T(\epsilon)$ is one of the primary objectives in a design process. Using the inequalities given above, the authors of [120] show that the following inequalities hold (again, the norms below are 1-norms):

$$\frac{z_{\min}}{z_{\max}} e^{-q_M T_1(\epsilon)} \le \epsilon = \frac{\|\mathbf{x}(T_1(\epsilon))\|}{\|\mathbf{x}_0\|} \le \frac{z_{\max}}{z_{\min}} e^{-q_m T_1(\epsilon)}, \tag{8.6}$$

for a certain delay $T_1(\epsilon) > 0$. This implies that

$$\frac{1}{q_M} \ln \frac{z_{\min}}{\epsilon z_{\max}} \le T_1(\epsilon) \le \frac{1}{q_m} \ln \frac{z_{\max}}{\epsilon z_{\min}}. \tag{8.7}$$

A simple numerical example is given to illustrate how these bounds are reasonably tight. The bounds obtained above are expected to provide useful information in a search for optimal parameters.

8.3 Conformal Mapping of Doubly Connected Regions

Solution of a large number of problems in modern technology, such as leakage of gas in a graphite brick of a gas-cooled nuclear reactor, analysis of stresses in solid propellant rocket grain, simultaneous flow of oil and gas in concentric pipes, and microwave theory, hinges critically on the possibility of conformal transformation of a doubly connected region into a circular annulus.

If D is a doubly connected region of the z-plane, then the frontier of D consists of two disjoint continua C_0 and C_1. It is well known [19] that D can be mapped conformally onto a circular annulus, in a one-to-one manner. Moreover, if a and b are the radii of two concentric circles of the annulus, then the modulus of D given by b/a is a number uniquely determined by D. The difficulties involved in finding such a mapping function and estimating the modulus of D are described by Kantorovich and Krylov [54]. In fact, studies concerning specific regions are very few in the literature. In this section, we review how the mapping problem of the region between a circle and a curvilinear polygon of n sides is reduced to an infinite system of linear algebraic equations, by a direct method. The truncated system of linear algebraic equations turns out to be strictly diagonally dominant.

Let the Jordan curves C_0 and C_1 bound externally and internally a doubly connected region D in the z-plane. Then the mapping function

$$w(z) = e^{[\log z + \phi(z)]}, \; z = x + iy = re^{i\theta},$$

which is unique except for an arbitrary rotation, maps $D + \partial D$ onto the annulus $0 \le a \le |w| \le b < \infty$, where the ratio b/a is unique and ϕ is regular in D. If ϕ has the form $\phi(z) = \sum_{-\infty}^{\infty} c_n z^n$, we then have

$$\log(z\bar{z}) + \phi(z) + \overline{\phi(z)} = \begin{cases} \log b^2, & \text{if } z \in C_0 \\ \log a^2, & \text{if } z \in C_1 \end{cases}$$

The requirement that ϕ satisfies the conditions given above is equivalent to solving the system of infinite linear equations:

$$\sum_{q=1}^{\infty} a_{pq} x_q = r_p, \; p = 1, 2, \ldots,$$

for suitable numbers a_{pq} and r_p (see Eq. (29) and the subsequent values in page 411, [105]). It can be shown that the truncated system of linear equations:

$$\sum_{q=1}^{n} a_{pq} x_q = r_p, \; p = 1, 2, \ldots, n$$

has the property that the determinant of the coefficient matrix is nonzero for all n. Thus the system has a unique solution for each n. Let $\mathbf{x}^{(n)}$ be the solution of the truncated system for each n. It then follows from a general principle (see, for instance, [54]) that $\lim_{n\to\infty} \mathbf{x}^{(n)}$ exists and is a solution of the infinite system. We refer to [105] for the details and numerical examples. We also refer to [102] for a similar procedure for the solution of the Poisson's equation describing a fluid flow problem.

8.4 Fluid Flow in Pipes

In this section, we consider the problem that arises from the idea that two fluids could be transported with one fluid inside a pipe of cross-section E and the other flowing in an annular domain D in the xy plane bounded internally by C_2 and externally by C_1. The flow velocity $w(x, y)$ satisfies the Poisson's equation:

$$w_{xx} + w_{yy} = -P/\mu \text{ in } D, \tag{8.8}$$

(P, μ being positive constants), with the boundary conditions:

$$w = 0 \text{ on } C_1 \text{ and } w = 0 \text{ on } C_2.$$

In this problem we are concerned with the rate of flow given by

$$R = \int\int w\, dxdy. \tag{8.9}$$

It can be shown that using the conformal mapping function

$$z = \frac{c}{1-\zeta},\ \zeta = \xi + i\eta, \tag{8.10}$$

we get

$$w = -\frac{P}{u\mu} z\bar{z} + \phi(z) + \overline{\phi(z)}. \tag{8.11}$$

We get an infinite series expression for w whose coefficients satisfy an infinite system of algebraic equations. These equations have been shown to have a unique solution. We refer to the work of Shivakumar, Chew, and Ji [102, 106] and [107] for the details. In these the following cases are considered

(a) C_0 and C_1 being eccentric circles.
(b) C_0 and C_1 being confocal ellipses.
(c) C_0 being a circle and C_1 being an ellipse.
(d) C_0 and C_1 being two ellipses.

Calculation of the rate of the flow suggests that the flow is a maximum when the inner boundary has the least perimeter and the outer boundary has the largest perimeter for a given area of flow.

We refer to the work of Luca, Kocabiyik and Shivakumar [68] for an application of the ideas given above in studying fluid flow in a pipe system where the inside of the outer pipe has a lining of porous media. This, in turn, has been shown to have applications in the cholesterol problem in arteries.

8.5 Mathieu Equation

Here we consider a Mathieu equation

$$\frac{d^2y}{dx^2} + (\lambda - 2q\cos 2x)y = 0, \tag{8.12}$$

for a given q with the boundary conditions: $y(0) = y(\pi/2) = 0$. Our interest is in the case when two consecutive eigenvalues merge and become equal for some values of the parameter q. This pair of merging points is called a *double point* for that value of q. It is well known that for real double points to occur, the parameter q must attain some pure imaginary value. In [115], Shivakumar and Xue developed an algorithm to compute some special double points. Theoretically, the method can achieve any required accuracy. We refer to [115] for the details. We briefly present the main results, here.

Using a solution of the form

$$y(x) = \sum_{r=1}^{\infty} x_r q^{-r} \sin 2rx,$$

the Mathieu equation is equivalent to the system of infinite linear algebraic equations given by $Bx = \lambda x$, where $B = (b_{ij})$ is an infinite tridiagonal matrix given by

$$b_{ij} = \begin{cases} -Q & \text{if } j = i - 1, \ i \geq 2 \\ 4i^2 & \text{if } j = i \\ 1 & \text{if } j = i + 1, \ i \geq 1. \end{cases}$$

Here $Q = -q^2 > 0$. Shivakumar and Xue show (Theorem 3.1, [115]) that there exists a unique double point in the interval $[4, 16]$. They also show that there is no double point in the interval $[16, 36]$, (Theorem 4.1, [115]). They present an algorithm for computing the double points. In fact, we have $\lambda \approx 11.20$ and $Q \approx 48.09$.

The problem of determining the bounds for the width of the instability intervals in the Mathieu equation has also been studied in the particular case when $q = h^2$, by Shivakumar and Ye [116]. We present the main result here. We consider the following boundary conditions: $y'(0) = y'(\pi/2) = 0$, and $y(0) = y'(\pi/2) = 0$, with the corresponding eigenvalues being denoted by $\{a_{2n}\}_0^\infty$ and $\{b_{2n}\}_1^\infty$, respectively. In this connection, the following inequalities are well known: $a_0 < b_1 < a_1 < b_2 < a_2 < b_3 < a_3 < \ldots$. The next result presents upper and lower bounds for $a_{2n} - b_{2n}$.

Theorem 8.5.1 (Theorem 2, [116]). *For* $n \geq \max\left\{\frac{h^2+1}{2}, 3\right\}$, *set*

$$p_n = \frac{8h^{4n}}{4^{2n}[(2n-1)!]^2}.$$

Then one has

$$\frac{p_n}{k_+}\left[1 - \frac{h^4}{8(2n-1-h^2)^2}\right] \leq a_{2n} - b_{2n} \leq \frac{p_n}{k_-},$$

for certain constants k_+ *and* k_-.

8.6 Bessel Functions

An iterative method developed by Shivakumar and Williams [112] was reviewed in Sect. 3.5 of Chap. 3. The authors also present (Example 5.2, [112]) an approximate method for determining the values of the Bessel functions $J_n(x)$ for all values of x lying in an interval. Recall that these functions satisfy the well-known recurrence relation:

$$J_{n+1}(x) = \frac{2n}{x} J_n(x) - J_{n-1}(x), \; n \in \mathbb{N}, \; 0 < x < 2.$$

Treating $J_0(x)$ as given and setting for $J_n(x) = x_n$, $n \in \mathbb{N}$, we get a system of equations whose coefficient matrix satisfies the conditions considered in Sect. 3.5. For instance, choosing $x = 0.6$, yields the following system of equations:

$$\begin{aligned} 10x_1 - 3x_2 &= 3J_0(0.6), \\ -3x_{n-1} + 10nx_n - 3x_{n+1} &= 0, \; n \geq 2, \end{aligned}$$

where $3J_0(0.6) \simeq 2.73601459$, [1]. From Corollary 3.5.1, Sect. 3.5, it follows that

$$\|\mathbf{x} - \mathbf{x}^{(p,n)}\|_\infty \leq \left\{ \frac{3}{7}(0.3)^{(p+1)} + \frac{9}{49(n+1)} \right\} J_0(0.6).$$

Thus for an error of less than 0.01, we choose $n = 16$ and $p = 6$. For the purpose of comparison, if we take $p = 16$, then, for instance, we have the following: $9.99555051\,E-07$ as against $9.9956\,E-07$ in [1], $9.99555051\,E-07$ as against $9.9956\,E-07$ in [1] for $J_6(0.6)$ and $1.61393490\,E-12$ as against $1.61396824\,E-12$ in [1] for $J_{10}(0.6)$. We refer to [112] for more details and comparison with known results.

8.7 Vibrating Membrane with a Hole

In this section, we review certain interesting results relating to the behavior of the minimal eigenvalue λ of the Dirichlet Laplacian in an annulus. Let D_1 be a disc on $\mathbb{R}^2$, with origin at the center of radius 1, $D_2 \subset D_1$ be a disc of radius $a < 1$, the center $(h, 0)$ of which is at a distance h from the origin. Let $\lambda(h)$ denote the minimum Dirichlet eigenvalue of the Laplacian in the annulus $D := D_h := D_1 \backslash D_2$. The following conjecture was proposed in [93]:

Conjecture. The minimal eigenvalue $\lambda(h)$ is a monotonic decreasing function of h on the interval $0 \leq h \leq (1 - a)$. In particular, $\lambda(0) > \lambda(h)$, $h > 0$.

The following results (Lemmas 1 and 2, [93]) were proved as supporting evidence to this conjecture:

Lemma 8.7.1. *We have*

$$\frac{d\lambda}{dh} = \int_S u_N^2 N_1 ds,$$

where N is the unit normal to $S = S_h$, pointing into the annulus D_h, N_1 is the projection of N onto x_1-axis, u_N is the normal derivative of u, and $u(x) = u(x_1, x_2)$ is the normalized $L^2(D)$ eigenfunction corresponding to the first eigenvalue λ.

Let $D(r)$ denote the disc $|x| \leq r$ and $\mu(r)$ be the first Dirichlet eigenvalue of the Laplacian in $D_1 \backslash D_r$. Then we have

Lemma 8.7.2. $\mu(a-h) < \lambda(h) < \mu(a+h)$, $0 < h < 1 - a$.

The conjecture was also substantiated by numerical results [93]. Subsequently, the authors prove the above conjecture in an electronic version of their article (https://www.math.ksu.edu/~ramm/papers/383.pdf). The argument of the proof is presented in the third paragraph in page 4. It is noteworthy to point out that this result has been now shown to be valid in all space dimensions by Kesavan [57].

8.8 Groundwater Flow

Here, we are concerned with the problem of finding the hydraulic head, ϕ in a nonhomogeneous porous medium, the region being bounded between two vertical impermeable boundaries, bounded on top by a sloping sinusoidal curve and unbounded in depth. The hydraulic conductivity K is modelled as $K(z) = e^{\beta z}$, supported by some data available from Atomic Energy of Canada Ltd. Here z is a real variable and $\beta \geq 0$. We recall a method that reduces the problem to that of solving an infinite system of linear equations. This method yields a Grammian matrix which is positive definite, and the truncation of this system yields an approximate solution that provides the best match with the given values on the top boundary. This is the work of Shivakumar, Williams, Ye and Ji [121], where further details are available. We only present a brief outline here.

We require ϕ to satisfy the equation:

$$\nabla.(e^{\beta z} \nabla \phi(x, z)) = 0,$$

where ∇ is the vector differential operator

$$\hat{i}\frac{\partial}{\partial x} + \hat{j}\frac{\partial}{\partial z}.$$

The domain under consideration is given by

$$0 < x < L, \quad -\infty < z < g(x) = -\left[\frac{ax}{L} + V\sin(\frac{2\pi nx}{L})\right], \tag{8.13}$$

where $L > 0, a \geq 0$ and V are constants and n is a positive integer. The boundary conditions are given by

$$\left.\frac{\partial\phi}{\partial x}\right|_{x=0} = \left.\frac{\partial\phi}{\partial x}\right|_{x=L} = 0, \tag{8.14}$$

$\phi(x,z)$ is bounded on $z \leq g(x)$, and $\phi(x,z) = z$ on $z = g(x)$. The determination of ϕ reduces to the problem of solving the infinite system of linear algebraic equations:

$$\sum_{m=0}^{\infty} b_{km}\alpha_m = c_k, \quad k = 0, 1, 2, \ldots, \tag{8.15}$$

where b_{km} are given by means of certain integrals. The infinite matrix $B = (b_{km})$ is the Grammian of a set of functions which arise in the study. The numbers b_{km} become difficult to evaluate for large values of k and m by numerical integration. The authors use an approach using modified Bessel functions, which gives analytical expressions for b_{km}. They also present numerical approximations and estimates for the error.

8.9 Eigenvalues of the Laplacian on an Elliptic Domain

The importance of eigenvalue problems concerning the Laplacian is well documented in classical and modern literature. Finding the eigenvalues for various geometries of the domains has posed many challenges for which the methods of approach include infinite systems of algebraic equations (as indicated in Sect. 8.5), asymptotic methods, integral equations, and finite element methods. Let us review the work of Shivakumar and Wu [114], where the details of earlier contributions are discussed. The eigenvalue problems of the Laplacian is represented by the Helmholtz equations, Telegraph equations or the equations of the vibrating membrane and is given by:

$$\frac{\partial^2 u}{\partial x^2} + \frac{\partial^2 u}{\partial y^2} + \lambda^2 u = 0 \text{ in } D, \quad u = 0 \text{ on } \partial D,$$

where D is a plane region bounded by a smooth curve ∂D. The eigenvalues k_n and corresponding eigenfunctions u_n describe the natural modes of vibration of the membrane. According to the maximum principle, k_n must be positive (for each n) for a nontrivial solution to exist. Further k_n, $n \in \mathbb{N}$ are ordered such that

$$0 < k_1 < k_2 < \cdots < k_n < \cdots .$$

The method described here provides a procedure to numerically calculate the eigenvalues.

Using complex variables $z = x + iy, \bar{z} = x - iy$, the problem becomes

$$\frac{\partial^2 u}{\partial z \partial \bar{z}} + \frac{\lambda^2}{4} u = 0 \text{ in } D \text{ and } \quad u = 0 \text{ on } C \tag{8.16}$$

with $u = u(z, \bar{z})$, It is well known that the general solution to (8.16) is given by

$$u = \left\{ f_0(z) - \int_0^z f_0(t) \frac{\partial}{\partial t} J_0 \left(\lambda \sqrt{\bar{z}(z-t)} \right) dt \right\} + \text{conjugate}, \tag{8.17}$$

where f_0 is an arbitrary holomorphic function which can be expressed as

$$f_0(z) = \sum_{n=0}^{\infty} a_n z^n \tag{8.18}$$

and J_0 is the Bessel function of the first kind of order 0, which is given by a series representation as,

$$J_0 \left(\lambda \sqrt{\bar{z}(z-t)} \right) = \sum_{k=0}^{\infty} \left(-\frac{\lambda^2}{4} \right)^k \frac{\bar{z}^k (z-t)^k}{k!\, k!}. \tag{8.19}$$

On substituting for f_0, we obtain the general solution to the Helmholtz equation as

$$\begin{aligned} u &= 2a_0 J_0(\lambda \sqrt{z\bar{z}}) \\ &\quad + \sum_{n=1}^{\infty} \sum_{k=0}^{\infty} A_{n,k} \left((z+\bar{z})^n + \sum_{m=1}^{n/2} \alpha_{m,n} (z+\bar{z})^{n-2m} (z\bar{z})^m a_n \right) (z\bar{z})^k, \end{aligned}$$

where $\alpha_{m,n} = (-1)^m \frac{n}{m} \binom{n-m-1}{m-1}$ and $A_{n,k}$ are constants determined in terms of certain Beta functions. The expression on the right-hand side demonstrates that the general solution of (8.16) without boundary conditions can be expressed in terms of powers of $z\bar{z}$ and $(z + \bar{z})$.

In our case, we consider the domain to be bounded by the ellipse represented by

$$\frac{x^2}{\alpha^2} + \frac{y^2}{\beta^2} = 1,$$

which can be expressed correspondingly in the complex plane by

$$(z+\bar{z})^2 = a + bz\bar{z}, \tag{8.20}$$

where $a = \dfrac{4\alpha^2\beta^2}{\beta^2-\alpha^2}$ and $b = \dfrac{4\alpha^2}{\alpha^2-\beta^2}$. After considerable manipulation, we get the value of u on the ellipse as,

$$\begin{aligned}
u &= 2a_0 + \sum_{n=1}^{\infty} A_{2n,0}b_{0,n}a_n \\
&+ \sum_{k=1}^{\infty}\left(\frac{\lambda^2}{4}\right)^k \frac{2a_0}{k!k!}(z\bar{z})^k \\
&+ \sum_{k=1}^{\infty}\left[\sum_{n=1}^{k}\left(A_{2n,k}b_{0,n} + \sum_{l=1}^{n} A_{2n,k-l}b_{l,n}\right)a_n\right](z\bar{z})^k \\
&+ \sum_{k=1}^{\infty}\left[\sum_{n=k+1}^{\infty}\left(A_{2n,k}b_{0,n} + \sum_{l=1}^{k} A_{2n,k-l}b_{l,n}\right)a_n\right](z\bar{z})^k,
\end{aligned} \tag{8.21}$$

for certain constants $b_{l,n}$ defined in terms of a and b. For $u = 0$ on the elliptic boundary, we equate the powers of $z\bar{z}$ to zero where we arrive at an infinite system of linear equations of the form

$$\sum_{k=0}^{\infty} d_{kn}a_n = 0, \quad n \in \mathbb{N}, \tag{8.22}$$

where d_{kn}'s are known polynomials of λ^2. In [114], the infinite system is truncated to an $n \times n$ system and numerical values are calculated and compared to existing results in the literature.

8.10 Shape of a Drum

A drumhead is conceived as a domain D in the plane whose boundary ∂D is clamped. It is well known that if a membrane D, held fixed along its boundary ∂D, is set in motion, its displacement obeys the wave equation

$$\frac{1}{2}\nabla^2 U + \lambda^2 U = 0, \ U = 0 \text{ on } \partial U.$$

Mark Kac in 1966 [52] published an interesting article on the question: Can one hear the shape of a drum? The phrasing of the title is due to Lipman Bers but the problem itself is older and can be traced back all the way to Hermann

Weyl. Similar questions can be asked for the Dirichlet problem for the Laplacian on domains in higher dimensions or on Riemannian manifolds, as well as for other elliptic differential operators such as the Cauchy–Riemann operator or the Dirac operator.

In 1964, John Milnor with the help of a result of Ernst Witt showed that there exist two Riemannian flat tori of dimension 16 with the same eigenvalues but different shapes. However, the problem in two dimensions remained open until 1992, when Gordon, Webb and Wolpert [39] found examples of distinct plane "drums" which "sound" the same. So, the answer to Kac's question is: for many shapes, one cannot hear the shape of the drum. However, some information can be inferred. Moreover, the answer is known to be "yes" for certain convex planar regions with analytic boundaries. A large number of mathematicians over four decades have contributed to the topic from various approaches, both theoretical and numerical.

In this section, we develop a constructive analytic approach to indicate how a pre-knowledge of the eigenvalues leads to the determination of the parameters of the boundary. This is based on a major contribution of Zelditch in 2000 [140], where a positive answer "yes" is given for certain regions with analytic boundaries. We apply this approach to a general boundary with biaxial symmetry, to a circle boundary, an ellipse boundary, and a square boundary. In the case of a square, we obtain an insight into why the analytical procedure does not, as expected, yield an answer. The work reported here is due to Shivakumar, Wu and Zhang [122].

Let D_L denote the class of bounded simply connected real analytic plain domains with reflection symmetries across two orthogonal axes, of which one has length L. Under generic conditions, if D_1 and D_2 are in D_L and if the Dirichlet spectra coincide then $D_1 = D_2$, up to a rigid motion [140].

So, mathematically the problem is, whether a pre-knowledge of the eigenvalues of the Laplacian in a region D leads to the identification of ∂D, the closed boundary of D. Specifically, we have

$$u_{xx} + u_{yy} + \lambda^2 u = 0 \text{ in } D, \tag{8.23}$$

$$u = 0 \text{ on } \partial D. \tag{8.24}$$

According to the maximum principle for linear elliptic partial differential equations, the infinitely many eigenvalues λ_n^2, $n \in \mathbb{N}$, are positive, real, ordered and satisfy

$$0 < \lambda_1^2 < \lambda_2^2 < \lambda_3^2 < \cdots < \lambda_n^2 < \cdots < \infty.$$

Using complex variables $z = x + iy$, $\bar{z} = x - iy$, Eqs. (8.23) and (8.24) become

$$u_{z\bar{z}} + \frac{\lambda^2}{4} u = 0 \text{ in } D, \tag{8.25}$$

$$u = 0 \text{ on } \partial D. \tag{8.26}$$

By Vekua [134], the completely integrated form of the solutions to the above equations are given by

$$u = \left\{ f_0(z) - \int_0^z f_0(t) \frac{\partial}{\partial t} J_0 \left(\lambda \sqrt{\bar{z}(z-t)} \right) dt \right\} + \text{conjugate}, \tag{8.27}$$

where $f_0(z)$ is an arbitrary holomorphic function which can be formally expressed as

$$f_0(z) = \sum_{n=0}^{\infty} a_n z^n,$$

and J_0 represents the Bessel function of first kind and order 0 given by

$$J_0 \left(\lambda \sqrt{\bar{z}(z-t)} \right) = \sum_{q=0}^{\infty} \left(-\frac{\lambda^2}{4} \right)^q \frac{\bar{z}^q (z-t)^q}{q! q!}. \tag{8.28}$$

Now we consider the problem for five types of boundaries: a general boundary with biaxial symmetry, a circular boundary, an elliptic boundary, a square boundary, and an annulus.

Before we study these cases, we recall an identity given in Abramowitz and Stegun [1], viz., when n is an even integer,

$$z^n + \bar{z}^n = \sum_{m=0}^{\frac{n}{2}} c_{m,n} (z + \bar{z})^{(n-2m)} (z\bar{z})^m, \tag{8.29}$$

for certain constants $c_{m,n}$. Now we consider the parametrized analytical boundary with biaxial symmetry to be given by

$$(z + \bar{z})^2 = \sum_{n=0}^{\infty} d_{n,1} (z\bar{z})^n \tag{8.30}$$

which yields, on using Cauchy products for infinite series,

$$(z + \bar{z})^{2p} = \sum_{n=0}^{\infty} d_{n,p} (z\bar{z})^n \tag{8.31}$$

where

$$d_{n,p} = \sum_{l=0}^{n} d_{l,p-1} d_{n-l,1}, \; p = 1, 2, \ldots.$$

For the specific problem at hand, we may assume that

$$f_0(z) = \sum_{n=0}^{\infty} a_{2n} z^{2n}$$

so that one has

$$u = 2a_0 J_0\left(\lambda\sqrt{z\bar{z}}\right) + \sum_{n=1}^{\infty} a_{2n} \sum_{k=0}^{\infty} \left(-\frac{\lambda^2}{4}\right)^k A_{2n,k}\left(z^{2n} + \bar{z}^{2n}\right)(z\bar{z})^k,$$

for certain constants $A_{2n,k}$. Upon substitution, this yields

$$u = 2a_0 J_0\left(\lambda\sqrt{z\bar{z}}\right) + \sum_{n=1}^{\infty} a_{2n} \sum_{k=0}^{\infty} -\left(\frac{\lambda^2}{4}\right)^k A_{2n,k} \sum_{m=0}^{n} \gamma_{m,n}(z+\bar{z})^{2(n-m)}(z\bar{z})^m,$$

where the constants $\gamma_{m,n}$ are given by $\gamma_{m,n} = (-1)^m \frac{2n(2n-m-1)!}{m!(2n-2m)!}$. After a rearrangement of summations, we get

$$\begin{aligned} u = {} & 2a_0 + \textstyle\sum_{n=1}^{\infty} a_{2n} D_{n,0,0,0} + \left[2a_0\left(-\frac{\lambda^2}{4}\right) + \sum_{n=1}^{\infty} a_{2n}\left\{\sum_{i=0}^{1}\sum_{p=0}^{1-i} D_{n,p,1-i-p,i}\right\}\right] z\bar{z} \\ & + \{\textstyle\sum_{q=2}^{\infty}\{\, 2a_0\left(-\frac{\lambda^2}{4}\right)^q \frac{1}{q!q!} + \sum_{n=1}^{q-1} a_{2n}\left[\sum_{i=0}^{n}\sum_{p=0}^{q-1} D_{n,p,q-i-p,i}\right] \\ & + \textstyle\sum_{n=q}^{\infty} a_{2n}\left[\sum_{i=0}^{q}\sum_{p=0}^{q-1} D_{n,p,q-i-p,i}\right]\}\}\,(z\bar{z})^q. \end{aligned}$$

Here, $D_{p,q,r,s}$ are certain constants. For a circular boundary given by $x^2 + y^2 = a^2$ we can consider the parametrization $z\bar{z} = a^2$. For an elliptic boundary given by

$$\frac{x^2}{\alpha^2} + \frac{y^2}{\beta^2} = 1$$

we can consider

$$(z+\bar{z})^2 = a + bz\bar{z},\ \alpha > \beta$$

where

$$a = \frac{4\alpha^2\beta^2}{\beta^2 - \alpha^2},\ b = \frac{4\alpha^2}{\alpha^2 - \beta^2}$$

and

$$\alpha^2 = \frac{a}{4-b},\ \beta^2 = -\frac{a}{b},\ a < 0, b > 0.$$

A square boundary given by $x = \pm a,\ y = \pm a$ can be parametrized as

$$z^4 + \bar{z}^4 = 2(z\bar{z})^2 - 16a^2(z\bar{z}) + 16a^4$$

or

$$z^2 + \bar{z}^2 = 4\,(z\bar{z} - 2a)^2\,.$$

For such a (square) boundary, the authors demonstrate why their analytical approach does not yield information of the boundary with sharp corners from a pre-knowledge of eigenvalues. When the boundary of the drum is an annulus we can hear the shape of the drum if the eigenvalues are known. In other words, for Eqs. (8.23) and (8.24), D is a totally connected region with the boundary conditions

$$u = 0 \text{ on } \pi_1 : x^2 + y^2 = a^2$$
$$u = 0 \text{ on } \pi_2 : x^2 + y^2 = b^2 \qquad a > b,$$

we can show that if the eigenvalues of λ are known, then the ratio of the annulus is uniquely determined.

8.11 On Zeros of Taylor Series

We wish to express the zeros z_n, $n \in \mathbb{N}$, of a Taylor series given by

$$y(x) = \sum_{n=0}^{\infty} c_n x^n = \prod_{n=1}^{\infty}(1 + a_n x), \tag{8.32}$$

where

$$c_0 = 1,\ c_1 = \sum_{k=1}^{\infty} a_k y'(0), \tag{8.33}$$

and $z_n = -\dfrac{1}{a_n}$. We assume that the zeros are positive and strictly increasing. Shivakumar and Zhang [117] show that for the second order linear differential equation [110]

$$y''(x) = f(x)y(x) \tag{8.34}$$

the formal Taylor series solution about $x = 0$ is given by

$$y(x) = y(0) + \frac{1}{2!}f(0)y(0)x^2 + \sum_{k=1}^{\infty} \frac{1}{(2k+2)!}[\sum_{s_1=0}^{1}\sum_{j=1}^{k+1} P_j(s_1, 2k, 0)]x^{2k+2}$$
$$+ y'(0)x + \frac{1}{3!}[f'(0)y(0) + f(0)y'(0)]x^3 + \sum_{k=1}^{\infty} \frac{1}{(2k+3)!}[\sum_{s_1=0}^{1}\sum_{j=1}^{k} P_j(s_1, 2k+1, 0)]x^{2k+3} \tag{8.35}$$

where $P_q(s_1, k, x)$ is given by

$$P_q(s_1, k, x) = \sum_{s_2=s_1}^{k-2(q-1)} \sum_{s_3=s_2}^{k-2(q-1)} \cdots \sum_{s_q=s_{q-1}}^{k-2(q-1)}$$
$$\binom{k}{s_q + 2(q-1)}\binom{s_q + 2(q-2)}{s_{q-1} + 2(q-2)}\binom{s_{q-1} + 2(q-3)}{s_{q-2} + 2(q-3)} \cdots \binom{s_3 + 2}{s_2 + 2}\binom{s_2}{s_1}$$
$$f^{(k-2(q-1)-s_q)}(x) f^{s_q - s_{q-1}}(x) f^{(s_{q-1}-s_{q-2})}(x) \ldots f^{(s_2-s_1)} y^{(s_1)}(x). \tag{8.36}$$

We note that

$$f(x) = \left(\sum_{n=1}^{\infty} \frac{a_n}{1 + a_n x}\right)^2 - \sum_{n=1}^{\infty} \frac{a_n^2}{(1 + a_n x)^2}$$

satisfies (8.34) where $y(x)$ is given by (8.32) and the coefficients c_n of the Taylor series are given in equation (8.35). To facilitate evaluation for c_n we need to find $f^p(0), p = 0, 1, 2, \ldots$. Using the notation

$$Q_p = \sum_{k=1}^{\infty} a_k^p, \; p = 1, 2, 3, \ldots,$$

we get

$$f(0) = Q_1^2 - Q_2,$$
$$f^p(0) = (-1)^{p+1} p! \left\{ \sum_{j=1}^{p+1} Q_j Q_{p+2-j} - (p+1) Q_{p+2} \right\}. \tag{8.37}$$

Rewrite (8.32) as

$$y(x) = \sum_{n=0}^{\infty} \frac{y^n(0)}{n!} x^n.$$

From (8.37), we note that $Q's$ can be expressed in terms of derivatives of $f(x)$ at $x = 0$ and using derivatives of $y(x)$ in (8.34) and comparing the coefficients of x^n in (8.35), one can obtain

$$Q_1 = d_1,$$

$$Q_2 = d_2,$$

$$\dots$$

$$Q_k = d_k,$$

where d_k's are functions of $c_1, c_2, \dots c_{k-1}$. For roots $z_n = -\frac{1}{a_n}$, a_n's satisfy the Vandermonde equation

$$\sum_{k=1}^{\infty} a_k^p = d_p.$$

Ran and Sereny [94] show that for a large class of infinite Vandermonde matrices the finite section method converges in the l^1 sense if the right-hand side of the equation is in a suitably weighted $l^1(\alpha)$ space.

References

1. Abramovitz, M., Stegun, I.: Handbook of Mathematical Functions. Dover, New York (1965)
2. Anderson, E.J., Nash, P.: Linear Programming in Infinite Dimensional Spaces. Wiley, New York (1987)
3. Arley, N., Borchsenius, V.: On the theory of infinite systems of differential equations and their application to the theory of stochastic processes and the perturbation theory of quantum mechanics. Acta Math. **76**, 261–322 (1945)
4. Baker, A.: Right eigenvalues for quaternionic matrices: a topological approach. Linear Algebra Appl. **286**, 303–309 (1999)
5. Barnett, S.: Leverrier's algorithm: a new proof and extensions. SIAM J. Matrix Anal. Appl. **10**, 551–556 (1989)
6. Beauwens, R.: Semistrict diagonal dominance. SIAM J. Numer. Anal. **13**, 109–112 (1976)
7. Bellman, R.: The boundedness of solutions of infinite systems of linear differential equations. Duke Math. J. **14**, 695–706 (1947)
8. Bellman, R.: Matrix Analysis. McGraw Hill, New York (1970)
9. Ben-Israel, A., Charnes, A.: An explicit solution of a special class of linear programming problems. Oper. Res. **16**, 1166–1175 (1968)
10. Ben-Israel, A., Greville, T.N.E.: Generalized Inverses: Theory and Applications, 2nd edn. Springer, New York (2003)
11. Berman, A., Plemmons, R.J.: Nonnegative Matrices in the Mathematical Sciences. Classics in Applied Mathematics, vol. 9. SIAM, Philadelphia, PA (1994). Revised reprint of the 1979 original
12. Bernkopf, M.: A history of infinite matrices. Arch. Hist. Exact Sci. **4**, 308–358 (1968)
13. Bhaskara Rao, K.P.S.: The Theory of Generalized Inverses over Commutative Rings. Algebra, Logic and Applications, vol. 17. Taylor & Francis Ltd, London (2002)
14. Brenner, J.L.: Matrices of quaternions. Pac. J. Math. **1**, 329–335 (1951)
15. Brenner, J.L.: A bound for a determinant with dominant main diagonal. Proc. Am. Math. Soc. **5**, 631–634 (1954)
16. Brenner, J.L.: Bounds for classical polynomials derivable by matrix methods. Proc. Am. Math. Soc. **30**, 353–362 (1971)
17. Brualdi, R.A.: Matrices, eigenvalues, and directed graphs. Linear Multilinear Algebra **11**, 143–165 (1982)
18. Campbell, S.L.: The Drazin inverse of an infinite matrix. SIAM J. Appl. Math. **31**, 492–503 (1976)
19. Caratheodory, C.: Theory of Functions of a Complex Variable, vol. 2. Chelsea Publishing Company, New York (1954)

P.N. Shivakumar et al., *Infinite Matrices and Their Recent Applications*,
DOI 10.1007/978-3-319-30180-8

20. Cooke, R.G., Infinite Matrices and Sequence Spaces. Dover, New York (1955)
21. Cottle, R.W., Pang, J-S., Stone, R.E.: The Linear Complementarity Problem. Academic, New York (1992)
22. Damiano, A., Gentili, G., Sreuppa, D.: Computations in the ring of quaternionic polynomials. J. Symb. Comput. **45**, 38–45 (2010)
23. Decell, H.P.: An application of the Cayley-Hamilton theorem to generalized matrix inversion. SIAM Rev. **7**, 526–528 (1965)
24. Drazin, M.P.: Pseudo-inverses in associative rings and semigroups. Am. Math. Mon. **65**, 506–514 (1958)
25. Drew, J., Johnson, C.R., van den Driessche, P., Strong forms of nonsingularity. Linear Algebra Appl. **162/164**, 187–204 (1992)
26. Faddeev, D.K., Faddeeva, W.N.: Computational Methods of Linear Algebra. Freeman, San Francisco (1963)
27. Faddeeva, V.N.: Computational Methods of Linear Algebra. Dover Books on Advanced Mathematics. Dover Publications, New York (1959)
28. Fan, K.: Inequalities for the sum of two M-matrices. In: Shisha, O. (ed.) Inequalities I, pp. 105–117. Academic, New York (1967)
29. Farid, F.O.: Spectral properties of perturbed linear operators and their application to infinite matrices. Proc. Am. Math. Soc. **112**, 1013–1022 (1991)
30. Farid, F.O.: Criteria for invertibility of diagonally dominant matrices. Linear Algebra Appl. **215**, 63–93 (1995)
31. Farid, F.O.: Notes on matrices with diagonally dominant properties. Linear Algebra Appl. **435**, 2793–2812 (2011)
32. Farid, F.O., Lancaster, P.: Spectral properties of diagonally dominant infinite matrices. I. Proc. R. Soc. Edinb. Sect. A **111**, 304–314 (1989)
33. Farid, F.O., Lancaster, P.: Spectral properties of diagonally dominant infinite matrices. II. Linear Algebra Appl. **143**, 7–17 (1991)
34. Fiedler, M., Ptak, V.: On matrices with non-positive off-diagonal elements and positive principal minors. Czech. J. Math. **12**, 382–400 (1962)
35. Gan, T.B., Huang, T.Z.: Simple criteria for nonsingular H-matrices. Linear Algebra Appl. **374**, 317–326 (2003)
36. Gil, M.I.: On invertibility and positive invertibility of matrices. Linear Algebra Appl. **327**, 95–104 (2001)
37. Giesbrecht, M., Labahn, G., Zhang, Y.: Computing Popov forms of matrices over PBW extensions. Computer Mathematics, pp. 61–66. Springer, New York (2014)
38. Gordon, B., Motzkin, T.S.: On the zeros of polynomials over division rings. Trans. Am. Math. Soc. **116**, 218–226 (1965)
39. Gordon, C., Webb, D.L., Wolpert, S.: One cannot hear the shape of a drum. Bull. Am. Math. Soc. **27**, 134–138 (1992)
40. Gowda, M.S., Tao, J.: Z-transformations on proper and symmetric cones: Z-transformations. Math. Program. Ser. B **117**, 195–221 (2009)
41. Groetsch, C.W.: Generalized Inverses of Linear Operators: Representation and Approximation. Monographs and Textbooks in Pure and Applied Mathematics, vol. 37. Marcel Dekker Inc, New York (1977)
42. Gu, W.: Generalized inverses of matrices of skew polynomials. Master thesis, University of Manitoba, Canada (2015)
43. Hettich, R., Kortanek, K.O.: Semi-infinite programming: theory, methods and applications, SIAM Rev. **35**, 380–429 (1993)
44. Horn, R.A., Johnson, C.R.: Matrix Analysis, 2nd edn. Cambridge University Press, Cambridge (2013)
45. Huang, L., Shivakurmar, P., Zhang, Y.: Solving quadratic quaternion equations via Maple. Can. Appl. Math. Q. **21**(2), 213–244 (2013)
46. Huang, L., Wang, Q., Zhang, Y.: The Moore-Penrose inverses of matrices over quaternion polynomial rings. Linear Algebra Appl. **475**, 45–61 (2015)

47. Huang, L., Zhang, Y.: Maple package for quaternion matrices, The University of Manitoba (2013)
48. Huang, T.Z., Xu, C.-X.: Generalized α-dominance. Comput. Math. Appl. **45**, 1721–1727 (2003)
49. Huang, T.Z., Zhu, Y.: Estimation of $\| A^{-1} \|_\infty$ for weakly chained diagonally dominant M-matrices. Linear Algebra Appl. **432**, 670–677 (2010)
50. Johnson, C.R., Tsatsomeros, M.J.: Convex sets of nonsingular and P-matrices. Linear Multilinear Algebra **38**, 233–239 (1995)
51. Jones, J., Karampetakis, N.P., Pugh, A.C.: The computation and application of the generalized inverse via Maple. J. Symb. Comput. **25**, 99–124 (1998)
52. Kac, M.: Can one hear the shape of a drum? Am. Math. Mon. **73**, 1–23 (1966)
53. Kanakadurga, K., Mercy Swarna, J., Sivakumar, K.C.: An algorithm for infinite linear programs. (Preprint)
54. Kantorovich, L.V., Krylov, V.I.: Approximate Methods of Higher Analysis. Interscience, New York (1964)
55. Kalauch, A.: Positive-off-diagonal operators on ordered normed spaces and maximum principles for M-operators, Ph.D. thesis, TU Dresden (2006)
56. Karamardian, S.: An existence theorem for the complementarity problem. J. Optim. Theory Appl. **19**, 227–232 (1976)
57. Kesavan, S.: On two functionals connected to the Laplacian in a class of doubly connected domains. Proc. R. Soc. Edinb. A **133**, 617–624 (2003)
58. Kostic, V.: On general principles of eigenvalue localizations via diagonal dominance. Adv. Comput. Math. **41**, 55–75 (2015)
59. Kulkarni, S.H., Sivakumar, K.C.: Applications of generalized inverses to interval linear programs in Hilbert spaces. Numer. Funct. Anal. Optim. **16**, 965–973 (1995)
60. Kulkarni, S.H., Sivakumar, K.C.: Explicit solutions of a special class of linear programming problems in Banach spaces. Acta Sci. Math. (Szeged) **62**, 457–465 (1996)
61. Kurmayya, T., Sivakumar, K.C.: Nonnegative Moore-Penrose inverses of Gram operators. Linear Algebra Appl. **422**, 471–476 (2007)
62. Lam, T.Y.: Lectures on Modules and Rings. Graduate Texts in Mathematics, vol. 180. Springer, New York (1999)
63. Lam, T.Y.: A First Course in Noncommutative Rings. Graduate Texts in Mathematics, vol. 131. Springer, New York (2001)
64. Lee, H.C.: Eigenvalues and canonical forms of matrices with quaternion coefficients. Proc. R. sIrish Acad. Sect. A **52**, 253–260 (1949)
65. Li, W.: The infinity norm bound for the inverse of nonsingular diagonal dominant matrices. Appl. Math. Lett. **21**, 258–263 (2008)
66. Li, B., Tsatsomeros, M.J.: Doubly diagonally dominant matrices. Linear Algebra Appl. **261**, 221–235 (1997)
67. Liu, S., Manganiello, F., Kschischang, F.R.: Kötter interpolation in skew polynomial rings. Des. Codes Cryptogr. Springer, New York (2013). doi:10.1007/s10623-012-9784-1
68. Luca, M., Kocabiyik, S., Shivakumar, P.N.: Flow past and through a porous medium in a doubly connected region. Appl. Math. Lett. **11**, 49–54 (1998)
69. Lyubich, Y.: Perron-Frobenius theory for Banach spaces with a hyperbolic cone. Integr. Equ. Oper. Theory **23**, 232–244 (1995)
70. Malejki, M.: Approximation and asymptotics of eigenvalues of unbounded self-adjoint Jacobi matrices acting in ℓ^2 by the use of finite submatrices. Cent. Eur. J. Math. **8**, 114–128 (2010)
71. Marek, I., Szyld, D.B.: Splittings of M-operators: irreducibility and the index of the iteration operator. Numer. Funct. Anal. Optim. **11**, 529–553 (1990)
72. Meek, D.S.: A new class of matrices with positive inverses. Linear Algebra Appl. **15**, 253–260 (1976)
73. Meyer, C.D. Jr.: Matrix Analysis and Applied Linear Algebra. SIAM, Philadelphia, PA (2000)
74. Muir, T.: A Treatise on the Theory of Determinants. Dover Phoenix Editions. Dover Publications, New York (2003)

75. Nabben, R.: Two-sided bounds on the inverses of diagonally dominant tridiagonal matrices. Linear Algebra Appl. **287**, 289–305 (1999)
76. Nabben, R.: Decay rates of the inverse of nonsymmetric tridiagonal and band matrices. SIAM J. Matrix Anal. Appl. **20**, 820–837 (1999) (electronic)
77. Nashed, M.Z. (ed.): Generalized Inverses and Applications. Academic, New York (1974)
78. Nashed, M.Z., Votruba, G.F.: A unified operator theory of generalized inverses. In: Nashed, M.Z. (ed.) Generalized Inverses and Applications, pp. 1–109. Academic, New York (1974)
79. Niven, I.: Equations in quaternions. Am. Math. Mon. **48**, 654–661 (1941)
80. Niven, I.: The roots of a quaternion. Am. Math. Mon. **49**, 386–388 (1942)
81. Opfer, G.: Polynomials and Vandermonde matrices over the field of quaternions. Electron. Trans. Numer. Anal. **36**, 9–16 (2009)
82. Ostrowski, A.M.: Note on bounds for determinants with dominant principal diagonal. Proc. Am. Math. Soc. **3**, 26–30 (1952)
83. Ostrowski, A.M.: On some conditions for nonvanishing of determinants. Proc. Am. Math. Soc. **12**, 268–273 (1961)
84. Peluso, R., Politi, T.: Some improvements for two-sided bounds on the inverse of diagonally dominant tridiagonal matrices. Linear Algebra Appl. **330**, 1–14 (2001)
85. Penrose, R.: A generalized inverse for matrices. Proc. Camb. Philos. Soc. **51**, 406–413 (1955)
86. Pereira, R., Rocha, P.: On the determinant of quaternionic polynomial matrices and its application to system stability. Math. Methods Appl. Sci. **31**, 99–122 (2008)
87. Pereira, R., Vettori, P.: Stability of Quaternionic linear systems. IEEE Trans. Autom. Control **51**, 518–523 (2006)
88. Pereira, R., Rocha, P., Vettori, P.: Algebraic tools for the study of quaternionic behavioral systems. Linear Algebra Appl. **400**, 121–140 (2005)
89. Petlovic, M.D., Stanimimirovic, P.S.: Symbolic computation of the Moore-Penrose inverse using a partitioning method. Int. J. Comput. Math. **82**(3), 355–367
90. Polak, E.: On the mathematical foundations of nondifferentiable optimization in engineering design. SIAM Rev. **29**, 21–89 (1987)
91. Puystjens, R.: Moore-Penrose inverses for matrices over some Noetherian rings. J. Pure Appl. Algebra **31**, 191–198 (1984)
92. Rajesh Kannan, M., Sivakumar, K.C.: On certain positivity classes of operators. Numer. Funct. Anal. Optim. **37**, 206–224 (2016)
93. Ramm , A.G., Shivakumar, P.N.: Inequalities for the minimal eigenvalue of the Laplacian in an annulus. Math. Inequal. Appl. **1**, 559–563 (1998)
94. Ran, A.M., Serény, A.: The finite section method for infinite Vandermonde matrices. Indag. Math. (N.S.) **23**, 884–899 (2012)
95. Rao, C.R., Mitra, S.K.: Generalized Inverse of Matrices and Its Applications. Wiley, New York (1971)
96. Roberts, M.L., Drazin, M.P.: Faithfully *-representing semigroups and groupoids with involution by infinite matrices. Linear Multilinear Algebra **60**, 511–524 (2012)
97. Rohn, J.: Inverse-positive interval matrices. Z. Angew. Math. Mech. **67**, 492–493 (1987)
98. Schaefer, H.H.: Some spectral properties of positive linear operators. Pac. J. Math. **10**, 1009–1019 (1960)
99. Schneider, H.: Olga Taussky-Todd's influence on matrix theory and matrix theorists. Linear Multilinear Algebra **5**, 197–224 (1977)
100. Shaw, L.: Solutions for infinite-matrix differential equations. J. Math. Anal. Appl. **41**, 373–383 (1973)
101. Shivakumar, P.N.: Diagonally dominant infinite matrices in linear equations. Util. Math. **1**, 235–248 (1972)
102. Shivakumar, P.N.: Viscous flow in pipes whose cross-sections are doubly connected regions. Appl. Sci. Res. **27**, 355–365 (1973)
103. Shivakumar, P.N., Chew, K.H.: A sufficient condition for nonvanishing of determinants. Proc. Am. Math. Soc. **43**, 63–66 (1974)

104. Shivakumar, P.N., Chew, K.H.: Iterations for diagonally dominant matrices. Can. Math. Bull. **19**, 375–377 (1976)
105. Shivakumar, P.N., Chew, K.H.: On the conformal mapping of a circular disc with a curvilinear polygonal hole. Proc. Indian Acad. Sci. **85**, 406–414 (1977)
106. Shivakumar, P.N., Chew, K.H.: On solutions of Poisson's equation for doubly connected regions, Pusat Pengajan Sains Matematik, University Sains Malaysia, Technical Report (1982), 2/82
107. Shivakumar, P.N., Ji, C.: On Poisson's equation for doubly connected regions. Can. Appl. Math. Q. **1**, 555–567 (1993)
108. Shivakumar, P.N., Ji, C.: Upper and lower bounds for inverse elements of finite and infinite tri-diagonal matrices. Linear Algebra Appl. **247**, 297–316 (1996)
109. Shivakumar, P.N., Ji, C.: On the nonsingularity of matrices with certain sign patterns. Linear Algebra Appl. **273**, 29–43. (1998)
110. Shivakumar, P.N., Leung, I.: On some higher order derivatives of solutions of $y'' + p(x)y = 0$. Util. Math. **41**, 3–10 (1992)
111. Shivakumar, P.N., Sivakumar, K.C.: A review of infinite matrices and their applications. Linear Algebra Appl. **430**, 976–998 (2009)
112. Shivakumar, P.N., Williams, J.J.: An iterative method with truncation for infinite linear systems. J. Comp. Appl. Math. **24**, 199–207 (1988)
113. Shivakumar, P.N., Wong, R.: Linear equations in infinite matrices. Linear Algebra Appl. **7**, 553–562 (1973)
114. Shivakumar, P.N., Wu, Y.: Eigenvalues of the Laplacian on an elliptic domain. Comput. Math. Appl. **55**, 1129–1136 (2008)
115. Shivakumar, P.N., Xue, J.: On the double points of a Mathieu equation. J. Comput. Appl. Math. **107**, 111–125 (1999)
116. Shivakumar, P.N., Ye, Q.: Bounds for the width of instability intervals in the Mathieu equation. In: Operator Theory: Advances and Applications, vol. 87, Birkhäuser, Basel pp. 348–357 (1996)
117. Shivakumar, P.N., Zhang, Y. On series solutions of $y'' - f(x)y = 0$ and applications. Adv. Differ. Equ. **47**, 6 pp. (2013)
118. Shivakumar, P.N., Chew, K.H., Williams, J.J.: Error bounds for the truncation of infinite linear differential systems. J. Inst. Math. Appl. **25**, 37–51 (1980)
119. Shivakumar, P.N., Williams, J.J., Rudraiah, N.: Eigenvalues for infinite matrices. Linear Algebra Appl. **96**, 35–63 (1987)
120. Shivakumar, P.N., Williams, J.J., Ye, Q., Marinov, C.: On two-sided bounds related to weakly diagonally dominant M-matrices with application to digital circuit dynamics. SIAM J. Matrix Anal. Appl. **17**, 298–312 (1996)
121. Shivakumar, P.N., Williams, J.J., Ye, Q., Ji, C.: An analysis of groundwater flow in an infinite region with a sinusoidal top. Numer. Funct. Anal. Optim. **21**, 263–271 (2000)
122. Shivakumar, P.N., Wu, Y., Zhang, Y.: Shape of a drum, a constructive approach. WSEAS Trans. Math. **10**, 21–31 (2011)
123. Sivakumar, K.C.: Moore-Penrose inverse of an invertible infinite matrix. Linear Multilinear Algebra **54**, 71–77 (2006)
124. Sivakumar, K.C.: Generalized inverses of an invertible infinite matrix. Linear Multilinear Algebra **54**, 113–122 (2006)
125. Sivakumar, K.C.: A new characterization of nonnegativity of Moore-Penrose inverses of Gram operators. Positivity **13**, 277–286 (2009)
126. Sivakumar, K.C., Weber, M.R.: On positive invertibility and splittings of operators in ordered Banach spaces. Vladikavkaz. Mat. Zh. **15**, 41–50 (2013)
127. Stanimirovic, P.S., Petkovic, M.D.: Computing generalized inverse of polynomial matrices by interpolation. Appl. Math. Comput. **172**, 508–523 (2006)
128. Tam, B.S., Yang, S., Zhang, X.: Invertibility of irreducible matrices. Linear Algebra Appl. **259**, 39–70 (1997)
129. Taussky, O.: A recurring theorem on determinants. Am. Math. Mon. **56**, 672–676 (1948)

130. Varah, J.M.: A lower bound for the smallest singular value of a matrix. Linear Algebra Appl. **11**, 3–5 (1975)
131. Varga, R.S.: Matrix Iterative Analysis. Prentice-Hall, Englewood Cliffs, NJ (1962)
132. Varga, R.S.: On recurring theorems on diagonal dominance. Linear Algebra Appl. **13**, 1–9 (1976)
133. Varga, R.S.: On diagonal dominance arguments for bounding $\| A^{-1} \|_{\infty}$. Linear Algebra Appl. **14**, 211–217 (1976)
134. Vekua, I.N.: New methods for solving elliptic equations [Translated from the Russian by D.E. Brown. Translation edited by A.B. Tayler]. North-Holland Series in Applied Mathematics and Mechanics, vol. 1. North-Holland/Wiley, Amsterdam/New York (1967)
135. von zur Gathen, J., Gerhard, J.: Modern Computer Algebra. Cambridge University Press, Cambridge (2013)
136. Vulikh, B.Z.: Introduction to the Theory of Cones in Normed Spaces (Russian). Izdat. Gosudarstv. Universitet Kalinin (1977)
137. Wang, F., Sun, D.-S., Zhao, J.-X.: New upper bounds for $\| A^{-1} \|_{\infty}$ of strictly diagonally dominant M-matrices. J. Inequal. Appl. **2015**, 172, 8 pp. (2015)
138. Williams, J.J., Ye, Q.: Infinite matrices bounded on weighted ℓ^1 spaces. Linear Algebra Appl. **438**, 4689–4700 (2013)
139. Yong, X.R.: Two properties of diagonally dominant matrices. Numer. Linear Algebra Appl. **3**, 173–177 (1996)
140. Zelditch, S.: Spectral determination of analytic bi-axisymmetric plane domains. Geom. Funct. Anal. **10**, 628–677 (2000)
141. Zhang, F.: Quaternions and matrices of quaternions. Linear Algebra Appl. **251**, 21–57 (1997)
142. Zhang, X., Gu, D.: A note on A. Brauer's theorem. Linear Algebra Appl. **196**, 163–174 (1994)
143. Zhang, Y.: A secret sharing scheme via skew polynomials. In: Proceedings of the 2010 International Conference on Computational Science and Its Applications, ICCSA'10, pp. 33–38. IEEE Computer Society, Washington, DC (2010)
144. Zhang, Y.: Computing Moore-Penrose inverses of Ore polynomials matrices. In: Hong, H., Yap, C. (eds.) ICMS 2014. Lecture Notes in Computer Science, vol. 8592, pp. 484–491. Springer, Berlin (2014)
145. Ziegler, M.: Quasi-optimal arithmetic for quaternion polynomials. In: Ibaraki, T., Katoh, N., Ono, H. (eds.) ISAAC 2003. Lecture Notes in Computer Science, vol. 2906, pp. 705–715. Springer, Berlin (2003)

Index

P.N. Shivakumar et al., *Infinite Matrices and Their Recent Applications*,
DOI 10.1007/978-3-319-30180-8

Printed in the United States
By Bookmasters